ENGINEERING PHYSICS

Dr. G. SHANMUGAM M.Sc., M.Phil., Ph.D

Associate Professor of Physics, Narayana Engineering College, Gudur-524 101, Andhra Pradesh

&

Former Assistant Professor, Department of Physics Vel Tech, Chennai-600 062, Tamilnadu

VASANTHA BOOKS PUBLISHERS
Chennai-600062

VASANTHA BOOKS PUBLISHERS
I-floor, 16/1-Poompozhil Nagar
Avadi, Chennai-600 062

ENGINEERING PHYSICS
First Edition 2019

Copyright © 2019 by VASANTHA BOOKS PUBLISHERS
All rights reserved. No part of this book may be reproduced or distributed or transmitted in any form or by any means, electronic, mechanical, photocopying, recording, or otherwise or stored in a database or retrieval system without the prior written permission of the publisher and author.

Typeset at VASANTHA BOOKS PUBLISHERS, 16/1, Poompozhil Nagar, Avadi, Chennai-600062

PREFACE

As per the New syllabus & Regulations 2017 prescribed by the Anna University, Chennai, this book **"ENGINEERING PHYSICS (PH8151)"** has been written by **Dr. G. SHANMUGAM**, Former Assistant Professor, Department of Physics, Vel Tech, Chennai-600062 for the first semester B.E/B. Tech degree course in all the branches

A sound understanding of the basic concepts of Physics and physical properties of materials is a major role for the future engineers and technologists to enable them to make a significant contribution in their fields of specialization. Since there is a very little time available for the teaching of Physics in an undergraduate engineering course, the book "Engineering Physics" is designed so as to avoid repetition of what the students have already learnt at the higher secondary level in school and to teach those topics which provide the essential background required for engineering subjects.

In recent years, pass percentage in Physics for the first year students is significantly low. Physics is the backbone of all engineering and technology but not the toughest subject for the students who one could not be able to pass. To guide the students (not only for passing the subject) for scoring more marks in Physics subject, I have taken lots of efforts to make this subject more interesting and easy-to-understand manner. The book has been presented in a very simple and lucid language.

This book contains five chapters and is tailored to conform to needs of engineering students at the undergraduate level. All the topics have been dealt with in simple and clear words to suit the requirements of students with diverse schooling backgrounds.

This book deals with the basics concepts of Physics that are of practical utility. It mainly focuses on the properties of matter, waves & optics and thermal physics and also covers topics on the quantum physics and crystal physics.

I hope that this book will be effectively useful for both students and teachers of Physics at various affiliated engineering colleges under Anna University. Suggestions, comments and feedback are most welcome for the improvement of this book.

Dr. G. Shanmugam M.Sc., M.Phil., Ph.D
Author

ACKNOWLEDGEMENTS

I express my sincere thanks to **Col. Prof. Vel. Dr. R. Rangarajan**, Founder Chancellor & President and **Dr. Sagunthala Rangarajan**, Foundress president, Vel Tech Group of Institutions for their moral support and encouragement for writing this book.

I also express my gratitude to **Dr. Mrs. Rangarajan Mahalakshmi. K,** Chair Person and Managing Trustee **Mr. K. V. D. Kishore Kumar,** Vice President, Vel Tech Group of Institutions for their kind support.

I wish to express my special thanks to **Dr. B. Nagalingeswara Raju,** Principal and **Dr. N. Edayadulla,** Associate Professor of Chemistry, Head of the Department of Science and Humanities for generously providing the time, support, help and encouragements, and relieving me of some of my teaching duties during the preparation of this book.

I have no words to thank, admire and appreciate the person behind this book **Prof. Dr. V. Krishnakumar**, Professor and Head, Department of Physics, Periyar University, Salem, my research supervisor, whose invaluable suggestions and incessant encouragement helped me to love the subject and understand it better. I am deeply indebted to him for his valuable advice and guidance during the preparation of this book.

I wish to place my hearty and sincere thanks to all the **Faculty members** of the Department of Physics, Vel Tech (owned by R. S. Trust), who encouraged me through moral support and ready to help me in all ways possible to complete this book successfully.

I thank **the lord Almighty,** the source of the knowledge who gave good health and mental courage to me during the entire span of time.

My sincere thanks to my mother **G. Vasantha** and Sister **G. Renukadevi** & beloved wife **S. Srimathi** and **S. Prashanna** for their prayers, affection and moral support throughout the period of book writing.

Dr. G. Shanmugam M.Sc., M.Phil., Ph.D
Author

SYLLABUS

UNIT I   Properties of Matter

Elasticity – Stress-strain diagram and its uses - factors affecting elastic modulus and tensile strength – torsional stress and deformations – twisting couple - torsion pendulum: theory and experiment - bending of beams - bending moment – cantilever: theory and experiment – uniform and non-uniform bending: theory and experiment - I-shaped girders - stress due to bending in beams.

UNIT II   Waves and Fiber Optics

Oscillatory motion – forced and damped oscillations: differential equation and its solution – plane progressive waves – wave equation. Lasers : population of energy levels, Einstein's A and B coefficients derivation – resonant cavity, optical amplification (qualitative) – Semiconductor lasers: homojunction and heterojunction – Fiber optics: principle, numerical aperture and acceptance angle - types of optical fibres (material, refractive index, mode) – losses associated with optical fibers - fibre optic sensors: pressure and displacement.

UNIT III   Thermal Physics

Transfer of heat energy – thermal expansion of solids and liquids – expansion joints - bimetallic strips - thermal conduction, convection and radiation – heat conductions in solids – thermal conductivity - Forbe's and Lee's disc method: theory and experiment - conduction through compound media (series and parallel) – thermal insulation – applications: heat exchangers, refrigerators, ovens and solar water heaters.

UNIT IV   Quantum Physics

Black body radiation – Planck's theory (derivation) – Compton effect: theory and experimental verification – wave particle duality – electron diffraction – concept of wave function and its physical significance – Schrödinger's wave equation – time independent and time dependent equations – particle in a one-dimensional rigid box – tunnelling (qualitative) - scanning tunnelling microscope.

UNIT V   Crystal Physics

Single crystalline, polycrystalline and amorphous materials – single crystals: unit cell, crystal systems, Bravais lattices, directions and planes in a crystal, Miller indices – inter-planar distances - coordination number and packing factor for SC, BCC, FCC, HCP and diamond structures - crystal imperfections: point defects, line defects – Burger vectors, stacking faults – role of imperfections in plastic deformation - growth of single crystals: solution and melt growth techniques.

CONTENTS

Unit-I	**Properties of Matter**	Page No
1.1	Introduction: Elasticity	1
1.2.	Stress-strain diagram and its uses	1
1.3	Factors affecting elastic modulus and tensile strength	2
1.4	Torsional stress and deformations	2
1.5	Twisting couple	3
1.6	Torsion pendulum: theory and experiment	5
1.7	Bending of beams	9
1.8	Bending moment	10
1.9	Cantilever: theory and experiment	11
1.10	Uniform bending: theory and experiment	15
1.11	Non-uniform bending: theory and experiment	17
1.12	I-shaped girders	19
1.13	Stress due to bending in beams.	

Unit-II	**Waves and Fiber Optics**	26
2.1	Introduction: oscillatory motion	26
2.2	Simple harmonic motion: Differential and its solution	26
2.3	Damped oscillations: differential equation and its solution	27
2.4	Forced oscillations: differential equation and its solution	29
2.5	Plane progressive waves and wave equation	32
2.6	Lasers : population of energy levels	34
2.7	Einstein's A and B coefficients derivation	37
2.8	Resonant cavity	41
2.9	Optical amplification	42
2.10	Semiconductor lasers	42
	2.10.1. Homojunction semiconductor laser	42
	2.10.2. Heterojunction semiconductor laser	44
2.11	Fiber optics	46
	2.11.1. Structure of optical fiber	46
	2.11.2. Expression for critical angle	47
2.12	Principle of light propagation in optical fibers	48
2.13	Numerical aperture	50
2.14	Types of optical fibers	50
2.15	Losses associated with optical fiber	53
2.16	Fiber optic sensors: pressure and displacement	55

Unit-III	**Thermal Physics**	62
3.1	Transfer of heat energy	62
3.2	Thermal expansion of solids	62
3.3	Thermal expansion of liquids	63
3.4	Expansion joints	63
3.5	Bimetallic strips	64
3.6	Three modes of transmission of heat	65
3.7	Heat conduction in solids	66
3.8	Thermal conductivity	66
3.9	Forbes method: Theory and experiment	66
3.10	Lee's disc method: Theory and experiment	70
3.11	Conduction through compound media (Series and parallel)	72
3.12	Thermal insulation	74
3.13	Applications	77
	3.13.1. Heat exchangers	77
	3.13.2. Refrigerators	80
	3.13.3. Ovens	83
	3.13.4. Solar water heaters	84
Unit-IV	**Quantum Physics**	92
4.1	Black body radiation	92
4.2	Planck's quantum theory	93
4.3	Compton effect:	96
	4.3.1. Theory and experimental verification	99
4.4	Wave particle duality	100
4.5	Electron diffraction	100
4.6	Concept of wave function and its physical significance	101
4.7	Schrödinger's wave equation	101
	4.7.1. Time dependent wave equation	101
	4.7.2. Time independent wave equation	103
4.8	Particle in a one-dimensional rigid box	105
4.9	Tunneling	109
4.10	Scanning tunneling microscope	109
Unit-V	**Crystal Physics**	116
5.1	Crystalline and non-crystalline materials	116
	5.1.1 Single polycrystalline materials	116
	5.1.2 Polycrystalline materials	116

5.2	Amorphous materials	116
5.3	Single crystals	116
	5.3.1. Unit cell	116
	5.3.2. Lattice parameters of the unit cell	116
	5.3.3. Crystal systems	117
	5.3.4. Bravais lattice	117
	5.3.5. Directions and planes in a crystal	117
5.4	Miller indices	118
5.5	Inter-planar distances	118
5.6	Simple cubic structure	120
5.7	Body centered cubic structure	122
5.8	Face centered cubic structure	124
5.9	Hexagonal closely packed structure	126
5.10	Diamond structure	130
5.11	Crystal imperfections	132
	5.11.1. Point defects	132
	5.11.2. Line defects	135
5.12	Burger vectors	137
5.13	Stacking faults	138
5.14	role of imperfections in plastic deformation	138
5.15	Growth of single crystals	139
	5.12.1. Solution growth	139
	5.12.2. Melt growth: Czochralski and Bridgmann methods	140

Unit-I
Properties of matter

1.1 Introduction: Elasticity

When an external force is applied on a material, the shape or size of a material is changed. After removing the force, the material regains its original shape and size. This phenomenon is known as elasticity of materials. The material which posses elastic property is called as elastic materials.

1.2. Stress-strain diagram and its uses

The elastic behavior of a material can be studied by plotting a curve between the stress along with Y-axis and the strain on the X-axis. The curve is called stress-strain curve. Let a wire be clamped at one end and loaded at the other end gradually from zero value until the wire breaks down. Fig. 1.1. shows the graph plotted between stress along Y-axis and strain along X-axis.

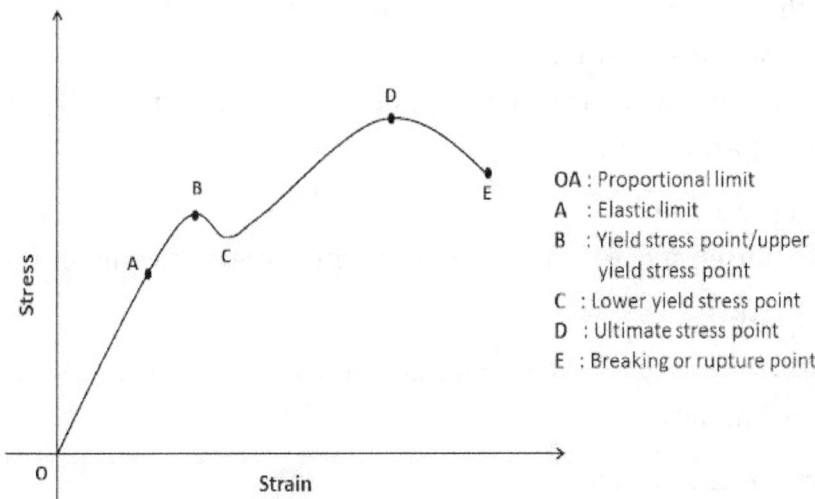

Fig. 1.1. stress and strain diagram

a). The part OA of the curve is a straight line which shows that up to a point 'A'. The point 'A' is called the proportional limit.

b). Point 'A' behaves as a perfectly elastic body and is known as elastic limit.

c). The material begins to deform without an increase of load at a point 'B' and is called yield point.

d). If the wire is further loaded, strain is reduced to a point 'C'. At the point 'C', the value of developed stress is large and called the ultimate tensile strength.

e). The stress corresponding to a point 'D', where the wire actually breaks down, is called the breaking stress.

Uses
- a). It is used to measure the elastic strength, yield strength and tensile strength of metals.
- b). It is used to estimate the working stress and safety factor.
- c). It is also used to identify the ductile and brittle materials.

1.3 Factors affecting elastic modulus and tensile strength

The following factors are affecting the elastic modulus and tensile strength.
- a). Stress
- b). Temperature
- c). Impurities
- d). Crystalline nature

a). Effect of stress
- When the material is subjected to a constant stress, it loses its elastic property within elastic limit.
- When the deformed material is allowed to rest sufficiently, it will regain its elastic property.

b). Effect of temperature
- Normally, the elastic property decreases with increase in temperature.

C). Effect of impurities
- The elastic property of a material may increase or decrease due to the addition of impurities.

d). Effect of crystalline nature
- For a given metal, the modulus of elasticity is more when it is in single crystal. In the polycrystalline state, its elastic modulus is comparatively small.

1.4 Torsional stress and deformations

The shear stress sets up in the shaft when equal and opposite torques are applied to the ends of a shaft about its axis is called torsional stress.

Shear strain does not only change with the amount of twist, but also it varies along the radial direction such that it is zero at the center and increases linearly towards the outer periphery.

1.5 Twisting couple on a wire

Consider a cylindrical wire of length (l) and radius (r) fixed at one end. It is twisted through an angle (θ) by applying couple to its lower end as shown in Fig. 1.2. Now the wire is said to be under torsion.

Let us consider one hollow cylinder of radius (x) and thickness (dx) as shown in diagram. AB is a line parallel to PQ on the surface of this cylinder. As the cylinder is twisted, the line AB is shifted to AC through an angle BAC = Φ.

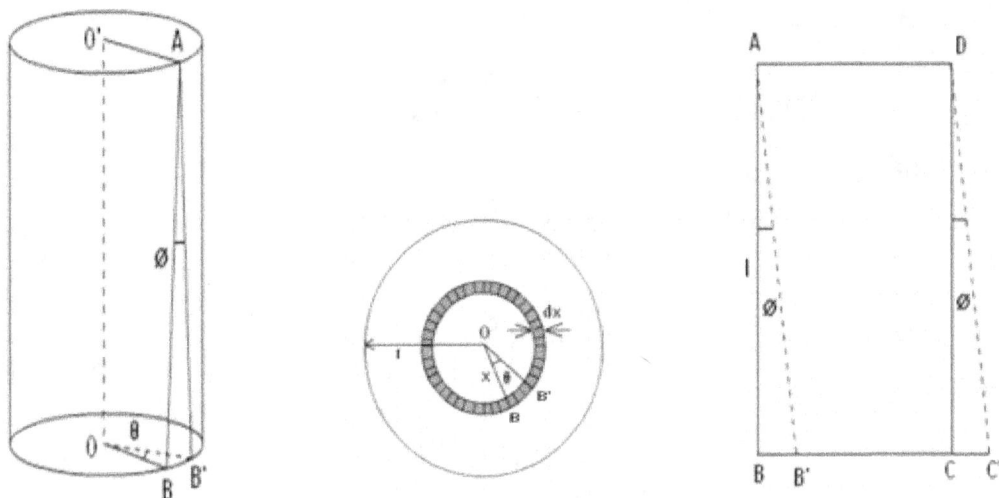

Fig. 1.2. Twisting couple on a wire

Shearing strain = Φ

Angle of twist at the free end = θ

From the figure, BC = xθ = lΦ

$$\Phi = \frac{x\theta}{l} \quad (1)$$

We know that shearing stress $(\eta) = \dfrac{\text{Shearing stress}}{\text{Shearing strain}}$

∴ Shearing stress = η × shearing strain = ηΦ (2)

Substitute equation (2) in equation (1), we get

$$\text{Shearing stress} = \frac{\eta x\theta}{l} \quad (3)$$

Shearing stress = $\dfrac{\text{Shearing force}}{\text{Area over which the force acts}}$ \hfill (4)

Area over which the force acts = $2\pi x dx$ \hfill (5)

Shearing force = Shearing stress × Area over which the force acts

Shearing force = $\dfrac{\eta x \theta}{l} 2\pi x dx$

Shearing force = $\dfrac{2\pi \eta \theta}{l} x^2 dx$ \hfill (6)

Moment of this force about

The axis PQ of the cylinder = Force × ⊥r distance

$$= \dfrac{2\pi \eta \theta}{l} x^2 dx \times x$$

$$= \dfrac{2\pi \eta \theta}{l} x^3 dx \hspace{2cm} (7)$$

Integrating equation (7) with limits x = 0, x = r

$$\text{Twisting couple (C)} = \int_0^r \dfrac{2\pi \eta \theta}{l} x^3 dx$$

$$C = \dfrac{\pi \eta \theta r^4}{2l} \hspace{2cm} (8)$$

∴ Twisting couple per unit twist (C) = $\dfrac{\pi \eta r^4}{2l}$, \hspace{2cm} If $\theta = 1$

∴ Twisting couple per unit twist of the hollow cylinder (C) = $\dfrac{\pi \eta}{2l}(r_2^4 - r_1^4)$

where r_1 and r_2 are inner and outer radius of cylinder.

Properties of matter

1.6 Torsional pendulum-Theory and experiments

A circular metallic disc suspended using a thin wire that executes torsional oscillation is called torsional pendulum.

A torsional pendulum consists of a metal wire suspended vertically with the upper end fixed. The lower end of the wire is connected to the centre of a metallic circular disc.

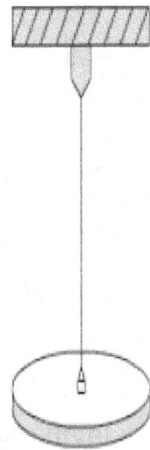

Fig. 1.3 Torsional pendulum

The restoring couple occurred in the wire by the rotation of a disc = $C\theta$ (1)

$$\text{Angular velocity of a disc} = \frac{d\theta}{dt}$$

$$\text{Applied couple} = I\frac{d^2\theta}{dt^2} \qquad (2)$$

In equilibrium, applied couple = restoring couple

$$I\frac{d^2\theta}{dt^2} = C\theta$$

$$\frac{d^2\theta}{dt^2} = \frac{C}{I}\theta \qquad (3)$$

Equation (3) represents the simple harmonic motion. Hence, the motion of the disc exhibits simple harmonic motion. Time period of the oscillation is given by

$$T = 2\pi \sqrt{\frac{\text{Displacement}}{\text{Acceleration}}}$$

$$T = 2\pi \sqrt{\frac{\theta}{\frac{C}{I} \times \theta}}$$

$$T = 2\pi \sqrt{\frac{I}{C}} \qquad (4)$$

Determination of rigidity modulus of the wire

A circular disc is suspended by a thin wire whose rigidity modulus is to be determined. The top end of the wire is fixed firmly in vertical support. The disc is rotated about its centre through a small angle and set it free. It executes torsional oscillations.

The disc is removed and its mass and diameter are measured. The time period of oscillation is

$$T = 2\pi \sqrt{\frac{I}{C}} \qquad (1)$$

Squaring equation (1) on both sides,

$$T^2 = \frac{4\pi^2 I}{C} \qquad (2)$$

We know that couple per unit twist (C)

$$C = \frac{\pi \eta r^4}{2l} \qquad (3)$$

Substitute equation (3) in equation (2), we get

$$T^2 = \frac{4\pi^2 I}{\frac{\pi \eta r^4}{2l}}$$

Properties of matter

The rigidity modulus of the material of the wire is given by

$$\eta = \frac{8\pi I}{r^4}\left(\frac{l}{T^2}\right) \quad (4)$$

where I = moment of inertia of disc = $\frac{MR^2}{2}$

Determination of moment of inertia of the disc using two equal symmetrical masses

The torsional pendulum consists of a steel or brass wire with one end fixed and the other end to the centre of a circular metallic disc as shown in Fig. 1.4

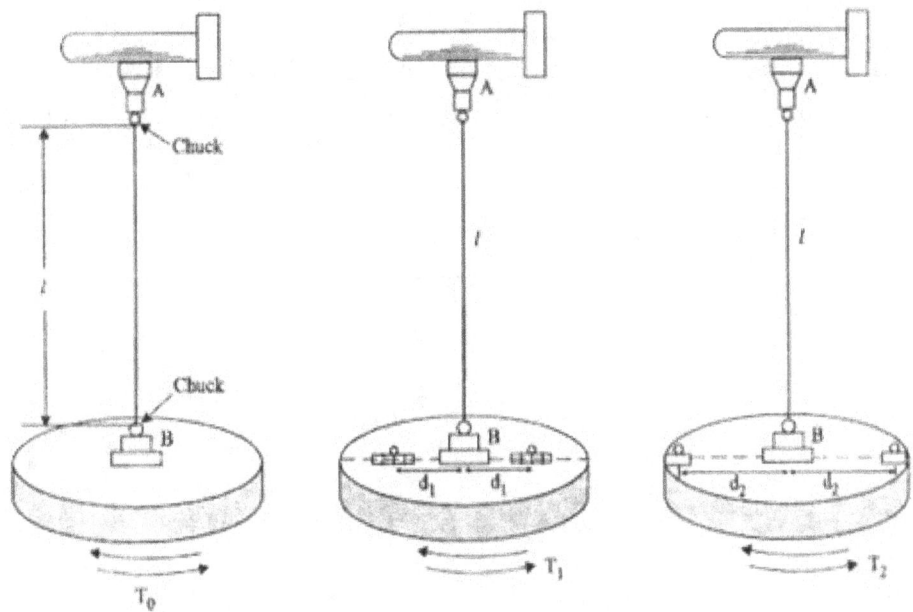

Fig. 1.4 Torsional oscillations without masses and with masses

When the disc is set into torsional oscillations without any cylindrical masses on the disc, the mean period of oscillation (T₀) is given by

$$T_0 = 2\pi\sqrt{\frac{I_0}{C}} \quad (1)$$

Where I_0 is moment of inertia of the disc about the axis of the wire.

Squaring equation (1) on both sides, we get

$$T_0^2 = \frac{4\pi^2 I_0}{C} \qquad (2)$$

When the disc is set into torsional oscillations with two equal cylindrical masses, the mean periods of oscillation are found as T_1 and T_2.

$$T_1 = 2\pi \sqrt{\frac{I_1}{C}} \qquad (3)$$

$$T_2 = 2\pi \sqrt{\frac{I_2}{C}} \qquad (4)$$

Where I_1 and I_2 are moment of inertia of the disc about the axis of the wire due to the cylindrical masses (m_1 and m_2). Squaring equations (3) and (4) on both sides, we get

$$T_1^2 = \frac{4\pi^2 I_1}{C} \qquad (5)$$

$$T_2^2 = \frac{4\pi^2 I_2}{C} \qquad (6)$$

substitute $I_1 = I_0 + 2i + 2md_1^2$ and $I_2 = I_0 + 2i + 2md_2^2$ in equations (5) and (6), we get

$$T_1^2 = \frac{4\pi^2}{C}(I_0 + 2i + 2md_1^2) \qquad (7)$$

$$T_2^2 = \frac{4\pi^2}{C}(I_0 + 2i + 2md_2^2) \qquad (8)$$

subtracting equation (7) from equation (8), we get

$$T_2^2 - T_1^2 = \frac{4\pi^2}{C} 2m(d_2^2 - d_1^2) \qquad (9)$$

Divide equation (2) by equation (8), we get

$$\frac{T_0^2}{T_2^2 - T_1^2} = \frac{I_0}{2m(d_2^2 - d_1^2)} \qquad (10)$$

$$I_0 = \frac{2m(d_2^2 - d_1^2)T_0^2}{T_2^2 - T_1^2} \qquad (11)$$

Equation (11) represents the moment of inertia of the disc about the axis of metallic disc

Determination of rigidity modulus of wire (suspension)

We know that restoring couple per unit twist

$$C = \frac{\pi \eta r^4}{2l} \qquad (12)$$

Substitute equation (12) in equation (9), we get

$$T_2^2 - T_1^2 = \frac{4\pi^2}{\frac{\pi \eta r^4}{2l}}[2m(d_2^2 - d_1^2)] \qquad (13)$$

$$T_2^2 - T_1^2 = \frac{16\pi m l(d_2^2 - d_1^2)}{\eta r^4}$$

$$\eta = \frac{16\pi m l(d_2^2 - d_1^2)}{(T_2^2 - T_1^2)r^4} \qquad (14)$$

Equation (14) represents the rigidity modulus of wire. Its unit is Nm^{-2}.

1.7 Bending of beam

a). Neutral surface and neutral axis:

In the middle of the beam, there is a layer which is not elongated or compressed due to bending of the beam. This layer is called neutral surface and the line at which the natural layer intersects the plane of bending is called neutral axis.

b). Plane of bending:

The plane in which bending takes place is known as plane of bending.

c). Neutral axis:

The line obtained by the intersection of neutral surface and plane of bending is called neutral axis.

d). Internal bending moment:

The resultant moments of all the internal couples are called internal bending moment.

1.8 Bending moment

The moment of the internal restoring couple is called bending moment.

Let us consider a beam ABCD having rectangular cross-section be bent in the form of an arc of a circle of radius 'R' with the centre at 'O' as shown in Fig.1.5. Consider a small portion 'ab'.

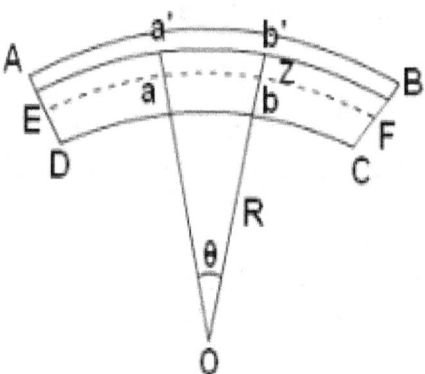

Fig. 1.5 Bending of beam

Before bending,	$a'b' = ab$
After bending,	$a'b' > ab$
When θ is small,	$a'b' = (R + Z)\theta$
	$ab = R\theta$
Increase in length	$a'b' - ab = Z\theta$

$$\text{Strain in } a'b' = \frac{Z\theta}{R\theta} = \frac{Z}{R} \quad (1)$$

Young's modulus of the beam is given by, $Y = \dfrac{\text{Stress}}{\text{Strain}}$

$$\therefore \text{Stress} = Y \times \text{strain} \qquad (2)$$

Substitute equation (1) in equation (2), we get that stress $= \dfrac{YZ}{R}$

we know that stress $= \dfrac{\text{Force}}{\text{Area}}$

Force = stress × area

$\therefore$ Total internal force acquired by the beam $= \dfrac{YZ}{R}\delta A$

we know that momentum = Force × distance

Momentum acquired by the beam $= \dfrac{YZ}{R}\delta A \times Z = \dfrac{Y}{R}\delta A Z^2$

Total momentum of the restoring couple per unit twist $= \dfrac{Y}{R}\sum \delta A Z^2$

Total momentum of the restoring couple per unit twist $= \dfrac{Y}{R}I_g \qquad (3)$

where $I_g = \sum \delta A Z^2$

1.9 Cantilever: theory and experiment

A beam which fixed horizontally at one end and loaded at the other end is called as cantilever.

Theory:

Let us consider a cantilever of length (l) fixed at the end 'A' and loaded at the free end 'B' by a weight 'W' as shown in Fig. 1.6. The end 'B' is depressed to B' and AB is the neutral axis. BB' represents the vertical depression at the free end.

Properties of matter

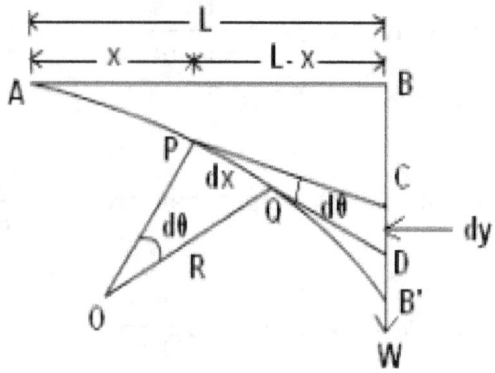

Fig. 1.6 Depression of cantilever

Consider the section of the cantilever 'P' at a distance 'x' from the fixed point 'A'. It is at a distance (*l*-x) from the loaded end B'. Considering the equilibrium of the portion PB'

$$\text{External bending moment} = W \times PB' = W(l - x) \qquad (1)$$

$$\text{Internal bending moment} = \frac{YI}{R} \qquad (2)$$

Under equilibrium condition,

External bending moment = Internal bending moment

$$W(l - x) = \frac{YI}{R} \qquad (3)$$

Q is another point at a distance 'dx' from P. i.e. PQ = dx

O is the centre of curvature of the arc PQ.

$$\therefore dx = R\,d\theta \qquad (4)$$

$$\text{Vertical depression}, CD = dy = (l - x)d\theta \qquad (5)$$

From equations (4) and (5), we have

$$R = \frac{(l - x)dx}{dy} \qquad (6)$$

Substitute equation (4) in equation (1), we get

$$dy = \frac{W}{YI}(l-x)^2 dx \qquad (7)$$

Total depression (y) of the beam at the free end is given by

$$\int dy = \int_0^l \frac{W}{YI}(l-x)^2 dx \qquad (8)$$

$$= \frac{W}{YI}\int_0^l (l^2 - 2lx - x^2) dx$$

$$= \frac{W}{YI}\left[l^2 x - \frac{2lx^2}{2} - \frac{x^3}{3}\right]_0^l$$

$$y = \frac{Wl^3}{3YI} \qquad (9)$$

Equation (9) represents the total depression of the beam at the free end.

∴ Young's modulus of the cantilever

$$Y = \frac{Wl^3}{3yI} \qquad (10)$$

For a beam of rectangular cross − section, $I = \frac{bd^3}{12}$ and $W = mg$

∴ Equation (10) can be written as

$$Y = \frac{4mgl^3}{bd^3 y} \text{ (Unit: Nm}^{-2}) \qquad (11)$$

Experiment:

The given bar is fixed at one end and a weight hanger is suspended at the other end as shown in Fig. 1.7. A pin is fixed vertically at the free end of the beam. A travelling microscope is used to focus on the pin. The initial reading in the microscope on the vertical scale is noted. A suitable mass (M) is placed on the hanger. The reading in the microscope is again noted. The difference between two readings of the microscope gives the depression 'y' corresponding to load 'M'.

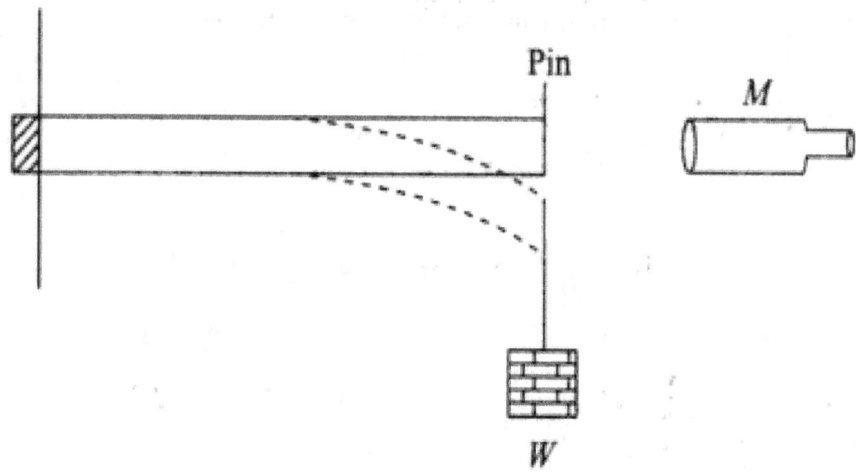

Fig. 1.7 Experimental set up for determining the Young's modulus of cantilever

The experiment is repeated by increasing the values of M in steps of 50 g. Then the experiment is also repeated by decreasing the weights.

Load	Microscope readings for depression (y)			Mean depression (y) for a load of M (kg)
	Load (increasing)	Load (decreasing)	Mean	
g	cm	cm	cm	cm
Mean (y) =				

On obtaining mean depression (y), length of beam (*l*), breadth (b) and thickness (d), Young's modulus of the cantilever (Y) can be obtained using the following equation:

$$Y = \frac{mgl^3}{bd^3y} \quad \text{(Unit: Nm}^{-2}\text{)}$$

Properties of matter

1.10 Uniform bending: Theory and experiment

If a beam is loaded uniformly on its both ends, if forms an arc of a circle. The elevation is produced in the beam. This type of bending is called as uniform bending.

Theory

Consider a beam AB arranged horizontally on two knife edges 'C' and 'D' symmetrically as shown in Fig. 1.8. so that AB = BD = a. The beam is loaded with equal weights 'W'.

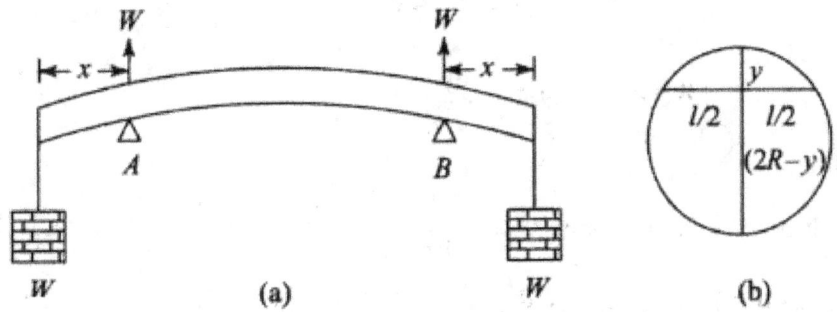

Fig. 1.8 (a) Uniform bending and (b) to find radius 'R'

$$\text{External bending moment} = Wa \qquad (1)$$

$$\text{Internal bending moment} = \frac{YI}{R} \qquad (2)$$

Under equilibrium condition,

External bending moment = Internal bending moment

$$Wa = \frac{YI}{R} \qquad (3)$$

Taking CD = l, y is the elevation of the mid-point (E) of the beam so that y = EF. From the Fig. 1.8(a), EF × EG = CE × ED (property of circle)

$$EF(2R - EF) = CE^2 \qquad \therefore CE = ED$$

$$y(2R - y) = \left(\frac{l}{2}\right)^2$$

$$2yR \quad y^2 = \frac{l^2}{4}$$

$$2yR = \frac{l^2}{4} \qquad (y^2 \text{ is negligible})$$

$$\frac{1}{R} = \frac{8y}{l^2} \qquad (4)$$

Substitute equation (2) in equation (1), we get

$$Wa = \frac{8y}{l^2} YI$$

Young's modulus (Y) of the beam, $Y = \dfrac{Wal^2}{8Iy}$ \qquad (5)

For a beam of rectangular cross-section,

$$I = \frac{bd^3}{12} \text{ and } W = mg$$

∴ Equation (10) can be written as

$$Y = \frac{3}{2}\frac{mgal^2}{bd^3 y} \qquad (6)$$

Experiment

A rectangular beam AB of uniform cross-section is supported horizontally on two knife edges K1 and K2 near the ends A and B as shown in Fig. 1.9.

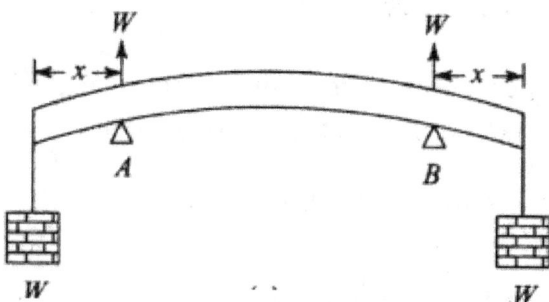

Fig. 1.9 Experimental setup for determining the Young's modulus of the beam by uniform bending

Properties of matter

Two weight hangers of equal masses are suspended from the ends of the beam. A pin is fixed vertically at the mid-point of the beam 'O'. A microscope is focused on the tip of the pin. Initial reading and final reading in the microscope on the vertical scale are noted.

Load	Microscope readings for elevation (y)			Mean elevation (y) for a load of M (kg)
	Load (increasing)	Load (decreasing)	Mean	
g	cm	cm	cm	cm
Mean (y) =				

On obtaining mean depression (y), length of beam (*l*), breadth (b) and thickness (d), Young's modulus of the cantilever (Y) can be obtained using the following equation:

$$Y = \frac{3}{2}\frac{mgal^2}{bd^3y} \text{ (Unit: Nm}^{-2}\text{)}$$

1.11 Non-uniform bending: (Theory and experiment)

If a beam is loaded at its mid-point, the depression is produced. This type of bending is called non-uniform bending.

Consider a uniform cross-section of a beam AB of length 'l' arranged horizontally on two knife edges K_1 and K_2 near the ends A and B as shown in Fig. 1.10.

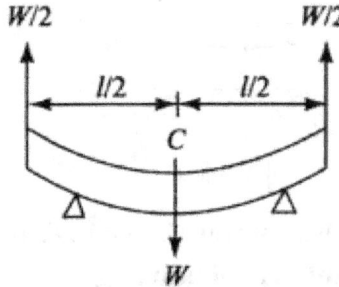

Fig. 1.10 Non-uniform bending

A weight (W) is applied at the midpoint 'O' of the beam. y is the depression at the midpoint 'O'. The bent beam is considered to be equivalent to two inverted cantilevers, fixed at 'O' with a weight (W/2).

Therefore, depression for a cantilever of length (l/2) and load (W/2) is given by

$$y = \frac{\left(\frac{W}{2}\right)\left(\frac{l}{2}\right)^3}{3IY} = \frac{\frac{W}{2}\frac{l^3}{8}}{3IY}$$

$$y = \frac{Wl^3}{48IY} \qquad (1)$$

$$Y = \frac{Wl^3}{48Iy} \qquad (2)$$

Equation (2) represents the Young's modulus of the beam under non-uniform bending.

For a beam of rectangular cross section, $I = \frac{bd^3}{12}$ and $W = mg$

Equation (2) can be written as

$$Y = \frac{mgl^3}{4bd^3y} \quad \text{(Unit: Nm}^{-2}\text{)} \qquad (3)$$

Experiment

A rectangular beam AB of uniform cross-section is supported horizontally on two knife edges K1 and K2 near the ends A and B as shown in Fig.1.11.

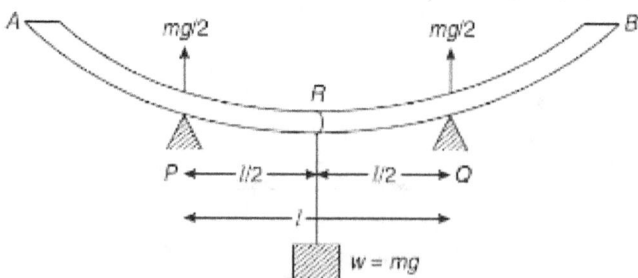

Fig. 1.9 Experimental setup for determining the Young's modulus of the beam by uniform bending

Two weight hangers of equal masses are suspended from the ends of the beam. A pin is fixed vertically at the mid-point of the beam 'O'. A microscope is focused on the tip of the pin. Initial reading and final reading in the microscope on the vertical scale are noted.

Load	Microscope readings for depression (y)			Mean depression (y) for a load of M (kg)
	Load (increasing)	Load (decreasing)	Mean	
g	cm	cm	cm	cm
		Mean (y) =		

On obtaining mean depression (y), length of beam (*l*), breadth (b) and thickness (d), Young's modulus of the cantilever (Y) can be obtained using the following equation:

$$Y = \frac{mgl^3}{4bd^3y} \text{ (Unit: Nm}^{-2}\text{)}$$

1.12 I-shape girders

The girders with upper and lower section broadened and the middle section trapped, so that it will have a shape of I and is called as I-shape girders.

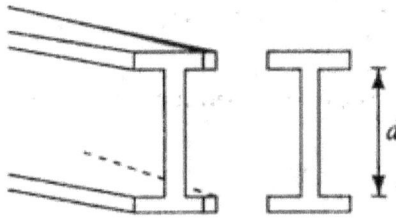

Fig. 1.10 I-shape girders

The depression of its mid-point is given by

$$Y = \frac{mgl^3}{4bd^3y} \text{ (Unit: Nm}^{-2}\text{)}$$

If it has rectangular cross-section of breadth 'b' and thickness 'd', then the depression is

$$y = \frac{mgl^3}{4bd^3Y}$$

When a beam is used as a girder, it should have minimum depression under its own weight. Small depression can be achieved by decreasing length (l) or increasing Y, b or d.

Advantages
    a). It provides a high bending moment.
    b). It has high Young's modulus.
    c). It has high durability.
    d). It is cheaper than the solid girders.

Applications
    a). It is used to produce iron rails.
    b). It is used in building construction.
    c). It supports frames and columns for trolley ways, lifts and hoists.
    d). It is used in machine bases.

Part-A TWO MARKS

1. **Define elasticity and plasticity.**

Sl. No	Elasticity	Plasticity
1.	When an applied force is removed, a material regains its original shape and size	When an applied force is removed, a material does not regain its original shape and size
2.	A materials possessed this property are known as elastic materials	A materials possessed this property are known as plastic materials
3.	Example: Rubber	Example: PVC

2. **Define stress and strain and write down their units.**

   Stress:

   It can be defined as force acting per unit area. Unit: Nm^{-1}

   Strain:

   It can be defined as the ratio of change in dimension and original dimension of a material. Unit: Nil.

3. **What are the different types of stress and strain?**

   Types of stress:
   a). Linear stress
   b). Volume or bulk stress
   c). Shearing stress

   Types of strain:
   a). Linear strain
   b). Volume or bulk strain
   c). Shearing strain

4. **Define elastic fatigue.**

   The temporary loss of elastic properties because of the action of repeated alternating deforming force is called elastic fatigue.

5. **State Hooke's law.**

   Within elastic limit, the stress developed in the body is directly proportional to strain produced in it.

   Stress α Strain

6. What do you infer from stress and strain diagram?

A graph plotted between strain along the X-axis and stress along Y-axis is known as stress-strain diagram. The elastic behavior of solid materials is studied by stress-strain diagram.

7. List out the uses of stress-strain diagram

a). It is used to estimate the elastic strength, yield strength and tensile strength of materials.

b). It is used to identify the ductile and brittle materials.

c). It is used estimate the working stress and safety factor of engineering materials

8. List the three modulii of elasticity.

a). Young's modulus

b). Bulk modulus

c). Rigidity modulus

9. What is Poisson's ratio?

The ratio of the lateral strain (β) and linear strain (α) is a constant for a given material. This constant is called as Poisson's ratio (σ).

$$\sigma = \frac{\beta}{\alpha}$$

10. List out the factors affecting elastic modulus and tensile strength.

a). Effect of stress

b). Effect of temperature

c). Effect of impurities

d). Effect of crystalline nature

11. How do temperature and impurity affect the elasticity of materials?

(i) Effect of temperature: Normally, the elastic property decreases with increase in temperature.

(ii) Effect of impurities: The elastic property of a material may increase or decrease due to the addition of impurities.

12. Define torque or moment of force.

It can defined as the product of the force (F) and the perpendicular distance (d) and is represented by (τ).

$$\tau = F \times d$$

Properties of matter

13. Define couple.

It is a pair of two equal and opposite forces acting on a body

14. Define torsional stress.

The shear stress sets up in the shaft when equal and opposite torques are applied to the ends of a shaft about its axis is called torsional stress.

15. Define internal bending moment or bending moment of a beam?

The resultants of the moments of all these internal couples are called internal bending moment or bending moment of a beam.

16. How are the various filaments of a beam affected when the beam is loaded?

A beam consists of a number of parallel layers placed one over the other and are called as filaments. Let the beam be subjected to deforming forces as its end. Due to the deforming force the beam bends. We know the beam consist of many filaments. Let us consider a centre filament at the beam. It is found that the filaments (layers) lying above centre filament gets elongated, while the filaments lying below centre filament gets compressed.

17. Define cantilever.

A beam which fixed horizontally at one end and loaded at the other end is called as cantilever.

18. When a wire is bent back and forth it becomes hot? Why?

When a wire is bent back and forth, it becomes hot, because of the friction developed in it. Not only a wire, even if you bent back and forth a metallic sheet also becomes hot due to the friction and displacement of the molecules.

19. An elastic wire is cut into half of its original length. How will it affect the maximum load the wire can support?

The breaking force is directly proportional to the cross-sectional area of the wire and not its length. Similarly, breaking stress is dependent only on the material of the wire and not its size. Thus, when an elastic wire is cut into half, there is no effect on the maximum load it can support.

Properties of matter

20. What is I-shape girder?

For the purpose of the stability, girders with the upper and lower parts of the cross-section will be broader and middle section is trapped, then the girders will have the shape of I, so it is called I-shape girders.

21. List out the advantages of I-shape girder
   a). It has high durability
   b). It is cheaper than the solid girder
   c). It has high bending moment.
   d). It has high young modulus

22. Why do we prefer I-shape girders rather than solid girders
   a). They are cheaper than the solid girders
   b). It would not undergo depression by its weight due to their low weight.
   c). Due to their larger depth, the depression produced is almost negligible which enhances the life of the girder

23. A cantilever of rectangular cross-section has a length of 50 cm. Its breadth is 3 cm and thickness 0.6 cm. A weight of 1 Kg is attached at the free end. The depression produced is 4.2 cm. Calculate Young's modulus of the material of the bar.

Given: l = 50 cm = 50×10⁻² m; b = 3 cm = 3×10⁻² m; d = 0.6 cm = 0.6×10⁻² m; y = 4.2 cm = 4.2×10⁻² m; M = 1 kg; g = 9.8 ms⁻²

$$\text{Young's modulus of the bar, Y} = \frac{4Mgl^3}{bd^3y}$$

$$Y = \frac{4 \times 1 \times 9.8 \times (50 \times 10^{-2})^3}{3 \times 10^{-2} \times (0.6 \times 10^{-2})^3 \times 4.2 \times 10^{-2}}$$

Y = 1.8 × 10¹⁰ Nm⁻²

24. Uniform rectangular bar 1 m long, 2 cm broad and 0.5 cm thick is supported on its flat face symmetrically on two knife edges 70 cm apart. If loads of 200 g are hung from the two ends the elevation of the centre of the bar is 48 mm. Find Young's modulus of the bar.

Given: a = 15×10⁻² m; M = 200 g = 200×10⁻³ kg; d = 0.5 cm = 0.5×10⁻² m
l = 70 cm = 70×10⁻² m; b = 2 cm = 2×10⁻² m; y = 48 mm = 48×10⁻³ m

Young's modulus of the bar, $Y = \dfrac{3}{2}\dfrac{Mgal^2}{bd^3y}$

$$Y = \dfrac{3}{2} \times \dfrac{200 \times 10^{-3} \times 9.8 \times 15 \times 10^{-2} \times (70 \times 10^{-2})^2}{2 \times 2 \times 10^{-2} \times (0.5 \times 10^{-2})^3 \times 48 \times 10^{-3}}$$

$Y = 1.8 \times 10^{10}$ Nm^{-2}

PART-B

1. Draw stress strain diagram and discuss the behaviour of ductile material under loading.
2. Explain the factors affecting the elasticity of the material.
3. Derive an expression for the torsional couple per unit angular twist when a cylinder is twisted.
4. What is torsional pendulum? How it is used to determine the a) moment of inertia of the disc and b) rigidity modulus of the wire using moment of inertia.
5. What is meant by bending moment of a beam? Derive an expression for the bending moment of a beam.
6. Derive the expression for the depression at the free end of a cantilever due to load. Describe an experiment to determine the young's modulus of the cantilever using this expression.
7. Explain with necessary theory the determination of young's modulus of elasticity of the material of the beam supported at its ends and loaded in the middle. Describe an experiment to determine the young's modulus of the material using this method.
8. How will you determine the young's modulus of material of a bar by non-uniform bending method?
9. Write a short note on I shaped girders. Give its applications and advantages.
10. Explain stress due to bending in beams.

Unit-II
Waves and Optics

2.1 Introduction: Oscillatory motion

It can be defined that a body moves to and fro repeatedly about a position.

Examples:
a). Motion of a pendulum
b). Oscillations of a loaded spring
c). To and fro motion of the prongs of tuning fiber.

2.2 Simple harmonic motion

It can be defined as that an oscillatory motion is harmonic if the displacement can be expressed in terms of sine and cosine function.

Characteristics:
a). The motion must be periodic
b). The motion should be oscillatory
c). The restoring force of simple harmonic oscillator is proportional to the displacement and its direction is always towards the mean position.
d). If there is no air resistance or friction, the motion started will continue indefinitely

Differential equation of Simple Harmonic Motion (SHM)

Consider a particle of mass 'm' executes SHM along a straight line. If 'y' is the displacement of particle at any time 't', then the restoring force 'F' is proportional to displacement 'y' and oppositively directed.
i.e. $F \alpha - y$

$$\therefore F = ky \qquad (1)$$

Where k is force constant and its unit is N/m.

If $a = \frac{d^2y}{dt^2}$ is acceleration at any instant, then the force is

$$F = m\frac{d^2y}{dt^2} \qquad (2)$$

Waves and Fiber Optics

From equations (1) and (2), $\quad m\dfrac{d^2y}{dt^2} = ky$

$$\therefore m\dfrac{d^2y}{dt^2} + ky = 0$$

$$\dfrac{d^2y}{dt^2} + \dfrac{k}{m}y = 0$$

$$\dfrac{d^2y}{dt^2} + \omega^2 y = 0 \tag{3}$$

where $\omega^2 = \dfrac{k}{m}$ is a constant.

Equation (3) represents the differential equation of simple harmonic motion. The solution of equation (3) is

$$Y = A\sin(\omega t + \phi)$$

Where A is an amplitude of the SHM and ϕ is initial phase.

2.3 Damped oscillations

Most of the oscillations in air or in any medium are damped. The amplitude of oscillations decreases with time and finally becomes zero. Such oscillations are called damped oscillations.

Examples
a). Oscillations of a pendulum
b). Electromagnetic oscillations in the tank circuits
c). Electromagnetic damping in galvanometer

Differential equation of damped oscillations and its solution

Let us consider a particle of mass 'm' attached to a spring executing damped SHM under resisting force and y be the displacement at any instant 't' as shown in Fig. 2.1.

This damped system is subjected to the following two forces

(a) Restoring force, $\quad\quad\quad\quad F_r = ky \tag{1}$

(b) Frictional force, $\quad\quad\quad\quad F_f = r\dfrac{dy}{dt} \tag{2}$

Total force is given by
$$F = F_r + F_f \tag{3}$$

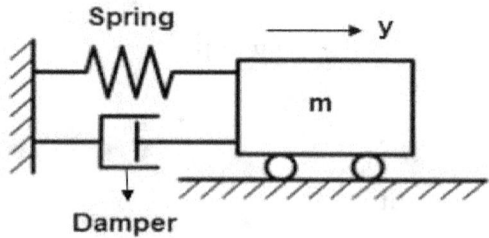

Fig. 2.1 Damped oscillations

Substitute equations (1) and (2) in equation (3), we get
$$F = -ky - r\frac{dy}{dt} \tag{4}$$

We know that
$$F = ma = m\frac{d^2y}{dt^2} \tag{5}$$

From equations (1) and (2),
$$m\frac{d^2y}{dt^2} = -ky - r\frac{dy}{dt}$$

$$\frac{d^2y}{dt^2} + \frac{r}{m}\frac{dy}{dt} + \frac{k}{m}y = 0$$

$$\frac{d^2y}{dt^2} + 2b\frac{dy}{dt} + \omega^2 y = 0 \tag{6}$$

Where $\frac{r}{m} = 2b$ and $\frac{k}{m} = \omega^2$ and b is known as damping factor.

Equation (6) is a differential equation of damped harmonic motion.

Let the solution of equation (6) be $\quad y = Ae^{\alpha t} \tag{7}$

Where A and α are arbitrary constants.

The general solution of equation (7) is given by

$$y = A_1 e^{(-b+\sqrt{b^2-\omega^2})t} + A_2 e^{(-b-\sqrt{b^2-\omega^2})t} \tag{8}$$

Where A_1 and A_2 are arbitrary constants and are determined from the boundary conditions.

Case (i) Heavy damping:

When $b^2 > \omega^2$, the powers $(b + \sqrt{b^2 - \omega^2})t$ and $(b - \sqrt{b^2 - \omega^2})t$ in equation (8) are negative. This displacement 'y' consists of two terms, both decreasing of exponentially to zero as shown in Fig. 2.2. This type of motion is known as over-damped or dead beat.

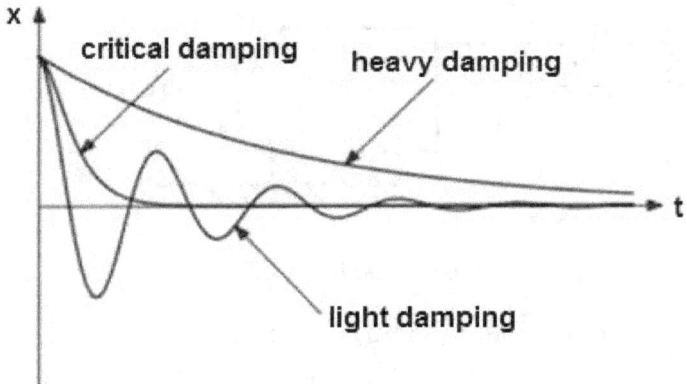

Fig. 2.2 Representation of damped oscillations

Case (ii) Critical damping:

When $b^2 = \omega^2, b^2 - \omega^2 = 0$, the equation (8) reduces to

$$y = (A_1 + A_2)e^{-bt} \qquad (8)$$

The above solution does not satisfy equation (6). Hence, equation (8) breaks down because of two coefficients become infinite. Such motion is called critically damped motion.

Case (iii) Light damping:

When $b^2 > \omega^2$, then $\sqrt{b^2 - \omega^2})$ is negative and imaginary. Equation (7) becomes,

$$y = ae^{-bt}\sin\left(\sqrt{\omega^2 - b^2}\,t + \phi\right) \qquad (9)$$

2.4 Forced oscillations:

When an object is oscillated by a periodic force of frequency other than its natural frequency of the body, then the oscillation is said to be forced oscillations.

Waves and Fiber Optics

Examples:
   a). Sound boards of stringed instruments
   b). By keeping the vibrating tuning fork

Differential equation and its solution of forced oscillations:

Consider a particle of mass ' m ' connected to a spring and it is driven by a periodic force.

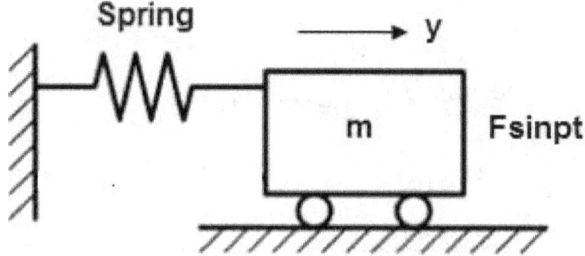

Fig. 2.3 Forced oscillations

The system is subjected to the following three forces.

(a) Restoring force, $\qquad F_r = \ ky \qquad$ (1)

(b) Frictional force, $\qquad F_f = \ r\dfrac{dy}{dt} \qquad$ (2)

(c) an external periodic force, $\qquad F_{ex} = Fsinpt \qquad$ (3)

Total force acting on the particle, $\quad F = F_r + F_f + F_{ex} \qquad$ (4)

Substituting equations (1), (2) & (3) in equation (4), we get

$$\therefore F = \ ky \quad r\dfrac{dy}{dt} + Fsinpt \qquad (5)$$

We know that $\qquad F = ma = m\dfrac{d^2y}{dt^2} \qquad$ (6)

From equations (1) and (2),

$$m\dfrac{d^2y}{dt^2} == \ ky \quad r\dfrac{dy}{dt} + Fsinpt$$

Waves and Fiber Optics

$$\frac{d^2y}{dt^2} + \frac{r}{m}\frac{dy}{dt} + \frac{k}{m}y = \frac{F}{m}\sin pt$$

$$\frac{d^2y}{dt^2} + 2b\frac{dy}{dt} + \omega^2 y = f\sin pt \qquad (7)$$

where $\frac{r}{m} = 2b, \frac{k}{m} = \omega^2$ and $\frac{F}{M} = f$

Equation (7) represents the differential equation of forced oscillation of the particle. The solution of differential equation (7) is,

$$y = A\sin(pt - \theta) \qquad (8)$$

Differentiating equation (8), we can obtain the following equations.

a). Amplitude of forced vibrations,

$$A = \frac{f}{\sqrt{(\omega^2 - p^2) + 4b^2 p^2}} \qquad (9)$$

b). Phase of forced oscillations,

$$\theta = \tan^{-1}\left(\frac{2bp}{\omega^2 - p^2}\right) \qquad (10)$$

Case (i)

When $p \ll \omega$, the amplitude of viberation is given by $A = \frac{f}{\omega^2}$ = constant and phase, $\theta = 0$. $\therefore$ The amplitude of vibration is independent of frequency of force.

Case (ii)

When $p = \omega$, the amplitude of viberation is given by $A = \frac{F}{r\omega}$ = constant and phase $(\theta) = \pi/2$. $\therefore$ The displacement lags behind the force by a phase of $\pi/2$.

Case (iii)

When $p \gg \omega$, the amplitude of viberation is given by $A = \frac{F}{mp^2}$ and phase $(\theta) = \pi$. $\therefore$ The displacement lags behind the force by a phase of π.

2.5 Plane progressive wave

The vibratory motion of a body is continuously transmitted in the same direction from one particle to the successive particle of the medium and travel forward through the medium due to its elasticity. These waves are called as plane progressive waves.

Wave equation of a plane progressive wave

Consider a plane progressive wave is propagating in a medium along positive x-axis. The position of particles O, A, B, C and D are shown in diagram. The curve joining these positions represents the progressive wave.

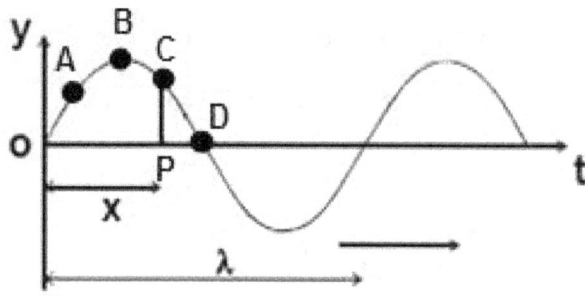

Fig. 2.4 Plane progressive waves

Let the particle begins to vibrate from origin ' O ' at time, t = 0. If y is displacement of the particle at time t, then the equation of particle is,

$$y = A \sin \omega t \qquad (1)$$

Where A is amplitude and ω is angular velocity.

substituting $\left(t - \dfrac{x}{v}\right)$ in place of 't' in equation (1),

The displacement of particle from O at time $\left(t - \dfrac{x}{v}\right)$ can be obtained as

$$y = A \sin \omega \left(t - \dfrac{x}{v}\right) \qquad (2)$$

$$y = A \sin \dfrac{2\pi}{T}\left(t - \dfrac{x}{v}\right) \qquad \text{where } \omega = \dfrac{2\pi}{T}$$

$$y = A \sin 2\pi \left(\dfrac{t}{T} - \dfrac{x}{vT}\right)$$

$$y = A\sin 2\pi\left(\frac{t}{T} - \frac{x}{\lambda}\right) \qquad \text{where } vT = \lambda$$

The above equation may also be expressed as

$$y = A\sin\frac{2\pi}{\lambda}\left(\frac{t\lambda}{T} - x\right)$$

$$y = A\sin\frac{2\pi}{\lambda}(vt - x) \qquad (3)$$

$$y = A\sin\left(\frac{2\pi vt}{\lambda} - \frac{2\pi x}{\lambda}\right)$$

$$y = A\sin 2\pi\left(\frac{\omega vt}{v} - \frac{\omega x}{v}\right)$$

∴ Equation (2) can be written as,

$$y = A\sin 2\pi\left(\omega t - \frac{\omega x}{v}\right)$$

$$y = A\sin 2\pi(\omega t - kx) \qquad (4)$$

where $k = \dfrac{\omega^2}{v}$

If ϕ be the phase difference between the wave travelling along x-positive axis and another wave, the equation of a wave may be expressed as

$$y = A\sin[(\omega t - kx) + \phi] \qquad (5)$$

Equation (5) represents the most general equation of a plane progressive wave.

Differential equation of wave motion

$$y = A\sin\frac{2\pi}{\lambda}(vt - x)$$

$$\frac{dy}{dt} = \frac{2\pi v A}{\lambda}\cos\frac{2\pi}{\lambda}(vt - x) \qquad (6)$$

$$\frac{dy}{dx} = \frac{2\pi A}{\lambda} \cos \frac{2\pi}{\lambda}(vt - x) \qquad (7)$$

$$\frac{dy}{dt} = v \frac{dy}{dx} \qquad (8)$$

$$\frac{d^2y}{dx^2} = A\left(\frac{2\pi}{\lambda}\right)^2 \sin \frac{2\pi}{\lambda}(vt - x) \qquad (9)$$

$$\frac{d^2y}{dx^2} = A\left(\frac{2\pi}{\lambda}\right)^2 v^2 \sin \frac{2\pi}{\lambda}(vt - x) \qquad (10)$$

Comparing equations (8) and (9), we get

$$\frac{d^2y}{dt^2} = v^2 \frac{d^2y}{dt^2} \qquad (11)$$

Equation (10) represents the differential equation of wave motion.

2.6 Laser

Laser stands for Light Amplification by Stimulated Emission of Radiation.

Characteristics of laser
  a). Monochromatic
  b). Coherent in nature
  c). Narrow beam
  d). Travel long distance without loss of intensity
  e). Amplify the light waves

Population of energy levels

It can be defined as that the number of atoms per unit volume in an energy level is known as population of energy levels.

Population inversion

It can be defined as that number of atoms in higher energy state (N_2) is more than number of atoms in lower energy state (N_1)

ie., $N_2 > N_1$

Pumping action

The process of supplying energy to the laser medium to achieve a state of population inversion is called as pumping action.

Methods for pumping action:
a). Optical pumping
b). Electrical discharge
c). Direct conversion
d). Inelastic collision between atoms

a). Optical pumping

When the photon energy of $h\upsilon$ strikes on the atoms in the lower energy state, it absorbs photon energy and moves to an excited energy state. This process is known as optical pumping and shown in Fig. 2.5.

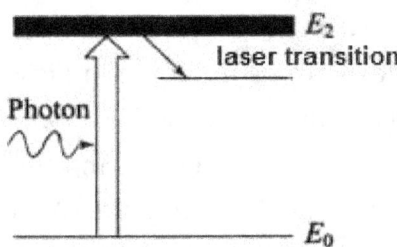

Fig. 2.5 Optical pumping

b). Electrical discharge

Electrons produced in an electrical discharge tube are accelerated to high velocities by a strong electric field. The accelerated electrons collide with the gas atoms.

During collision, the energy of the electrons is transferred to gas atoms. Thereby atoms gain energy and move to excited state. This process is called as electrical discharge.

Energy is represented by the following equation

$$A + e^* \rightarrow A^* + e$$

Where A – gas atom in ground state
      A* - gas atom in excited energy state
      e* - electrons with more kinetic energy
      e – some electrons with less energy

c). Direct conversion:

When an electrical energy is applied to a semiconductor, the recombination of electrons and holes take place. During the recombination process, electrical energy is directly converted into light energy. The direct conversion process is shown in Fig. 2.6.

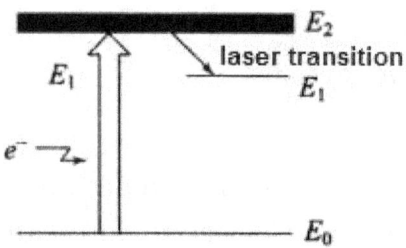

Fig. 2.6 Direct conversion process

d). Inelastic collision between atoms

Inelastic collision between atoms is shown in Fig. 2.7. In this method, a combination of two gases (say A and B) is used. The excited energy levels of gases of A and B nearly coincides each other. During the electrical discharge, atoms of gas 'A' are excited to higher energy state 'A*' due to collision with electrons.

$$A + e \rightarrow A + e$$

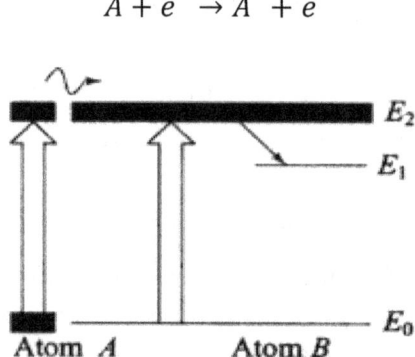

Fig. 2.7 Inelastic collision between atoms

Now A* atoms at higher energy state collide with 'B' atoms in lower energy state. Due to this inelastic collision, 'B' atoms gain energy and excited to higher energy state 'B*'. Hence, A atoms lose energy and return to lower energy state.

$$A + B \rightarrow A + e$$

2.7 Einstein's A and B coefficients derivation

Consider an atom that has only two energy levels, E_1 and E_2. When it is exposed to light radiation, three distinct processes can take place. (i) Stimulated absorption, (ii) Spontaneous emission and (iii) Stimulated emission.

(i) Stimulated absorption:

An atom or molecule in the ground state E_1 can absorbs a photon of energy υ and go to the higher energy state E_2. This process is known as stimulated absorption and is illustrated in Fig. 2.8.

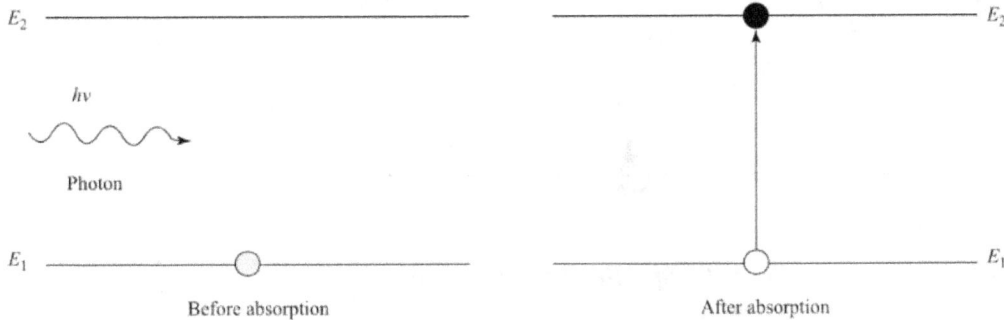

Before absorption   After absorption

Fig. 2.8 Stimulated absorption

The rate of absorption R_{ab} is directly proportional to the number of atoms N_1 in the ground state and to the energy density of light radiation ρ. ie.,

$$R_{ab} \alpha\ N_1\ \rho$$

$$R_{ab} = B_{12}\ N_1\ \rho \qquad (1)$$

Where B_{12} is the probability of absorption per unit time.

Normally, the excited atoms in the higher energy state E_2 will make a transition back to the lower energy state with the emission of a photon. Such an emission can take place by the two methods given below.

(ii) Spontaneous emission:

In spontaneous emission, the atoms in the higher energy state E_2 return to the ground state by emitting the photon of energy spontaneously. This process is independent of external radiation. The rate of the spontaneous emission R_{sp} is directly proportional to the number of atoms in higher energy state N_2.

$$R_{sp} \alpha\ N_2$$

$$R_{sp} = A_{21}\ N_2 \qquad (2)$$

Where A_{21} is the probability per unit time that atoms will spontaneously fall to the ground state without any external radiation. This process is illustrated in Fig. 2.9.

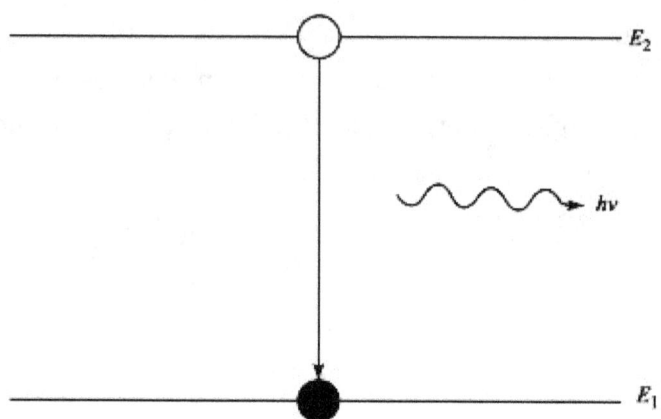

Fig. 2.9 Spontaneous emission

(iii) Stimulated emission:

In stimulated emission, a photon having energy υ stimulates an atom in the higher energy state to make a transition to the lower state with the creation of a second photon, as illustrated in Fig. 2.10.

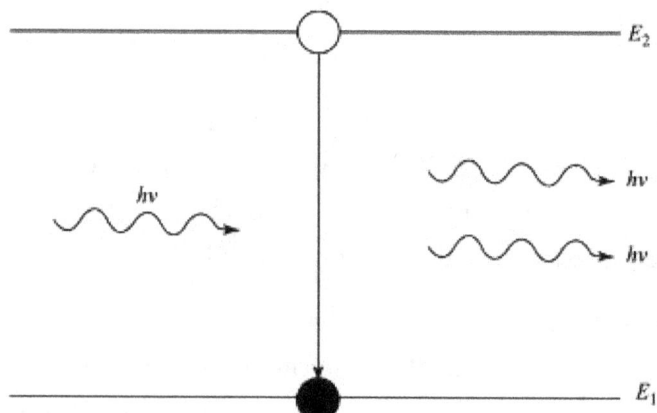

Fig. 2.10 Stimulated emission

The rate of stimulated emission R_{sp} is directly proportional to the number atoms in the higher energy state E_2 and the energy density of light radiation. ie.,

Waves and Fiber Optics

$$R_{st} \propto N_2 \rho$$

$$R_{st} = B_{21} N_2 \rho \qquad (3)$$

Where B_{21} is probability per unit time that the atoms undergo transition from higher energy state E_2 to the ground energy state E_1 by the stimulated emission.

At thermal equilibrium condition, the rate of absorption = the rate of emission. ie.,

$$R_{ab} = R_{sp} + R_{st}$$

$$B_{12} N_1 \rho = A_{21} N_2 + B_{21} N_2 \rho$$

$$B_{12} N_1 \rho - B_{21} N_2 \rho = A_{21} N_2$$

$$(B_{12} N_1 - B_{21} N_2)\rho = A_{21} N_2$$

$$\rho = \frac{A_{21} N_2}{B_{12} N_1 - B_{21} N_2} \qquad (4)$$

Divide each and every term on the RHS of equation (4) by N_2, we get

$$\rho = \frac{\frac{A_{21} N_2}{N_2}}{\frac{B_{12} N_1 - B_{21} N_2}{N_2}}$$

$$\rho = \frac{A_{21}}{B_{12}\left(\frac{N_1}{N_2}\right) - B_{21}} \qquad (5)$$

According to the Boltzmann's distribution function, the distribution of atoms in energy states can be written as

$$\frac{N_1}{N_2} = \frac{e^{-E_1/kT}}{e^{-E_2/kT}} = e^{(E_2-E_1)/kT} \qquad (6)$$

Substituting equation (6) in equation (5), we get

$$\rho = \frac{A_{21}}{B_{12}\left[e^{(E_2-E_1)/kT}\right] - B_{21}}$$

$$\rho = \frac{A_{21}}{B_{12}e^{h\nu/kT} - B_{21}}$$

$$\rho = \frac{\frac{A_{21}}{B_{21}}}{\left(\frac{B_{12}}{B_{21}}\right)e^{h\nu/kT} - 1} \tag{7}$$

Where $E_2 - E_1 = h\nu$. The coefficients A_{21}, B_{12} and B_{21} are known as Einstein's coefficients.

According to the Planck's quantum theory for black body radiation, the energy density is given by

$$\rho = \frac{8\pi h\nu^3}{c^3}\frac{1}{e^{h\nu/kT} - 1} \tag{8}$$

Comparing equation (7) with equation (8), we get

$$B_{12} = B_{21}$$

$$\frac{A_{21}}{B_{21}} = \frac{8\pi h\nu^3}{c^3}$$

The ratio of stimulated emission to the spontaneous emission is given by

$$\frac{R_{st}}{R_{sp}} = \frac{B_{21} N_2 \rho}{A_{21} N_2} = \frac{1}{e^{h\nu/kT} - 1} \tag{9}$$

From equation (9), Einstein proved that the existence of stimulated emission of radiation which is predominant over the spontaneous emission and absorption.

2.8 Resonant Cavity

A resonator consists of a pair of mirrors, plane or spherical having a common principal axis. Light beam passing through an active medium with population inversion is amplified. At the right end, amplified light is ready to be transmitted as laser. The structure of resonant cavity is shown in Fig. 2.11.

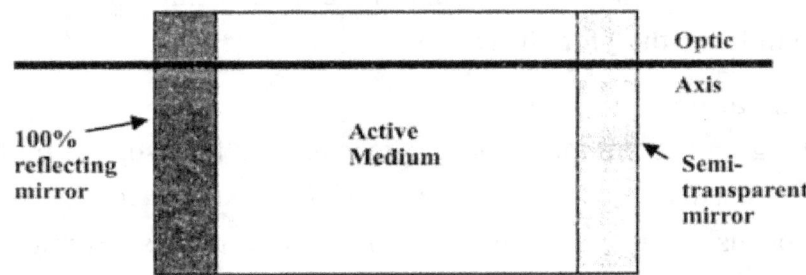

Fig. 2.11 Structure of resonant cavity

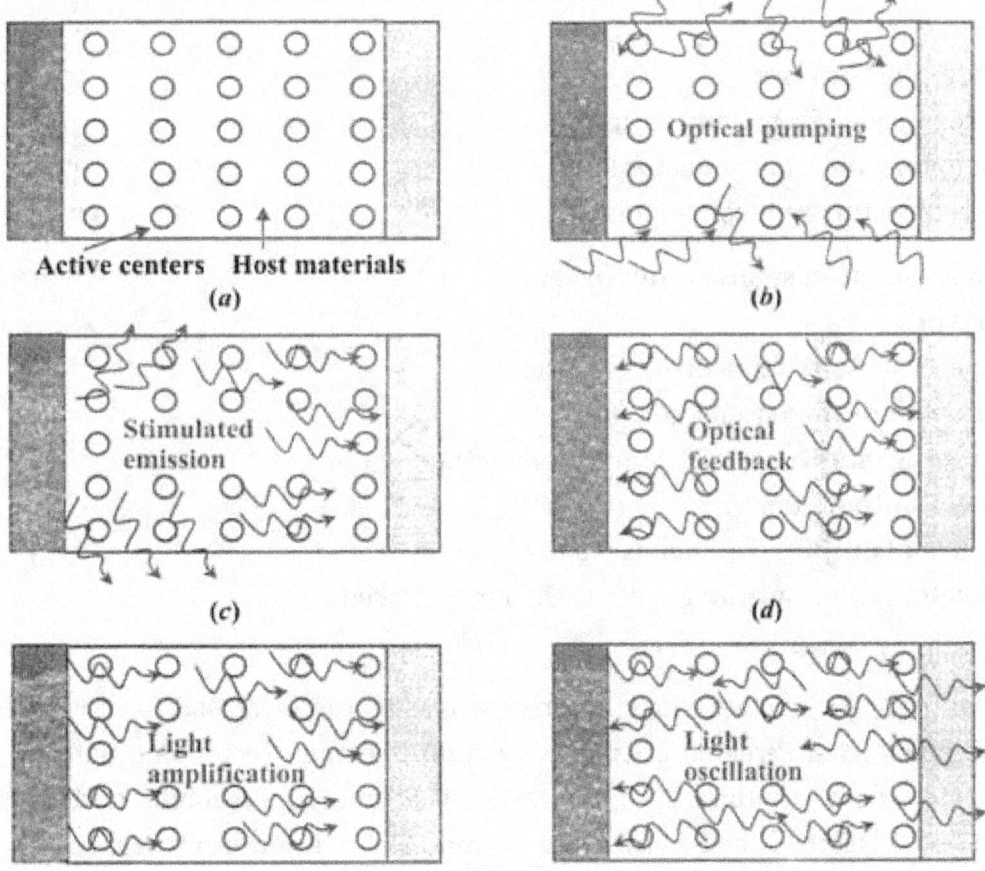

Fig. 2.12 Laser action in resonant cavity

In order to increase the light intensity, part of the output light has to be sent back into the active medium. This can be achieved by placing the active medium between two mirrors (resonator) which reflect part of the output energy back to active medium.

The reflected part of the output beam at other end stimulates the excited atoms to make a transition to ground state by emitting photons. The intensity of the light beam therefore increases.

2.9 Optical amplification

On switching on the pumping source, light energy is absorbed by the atoms which are raised to a higher energy level (metastable state). If the pumping radiation is adequate, population inversion occurs. Stimulated emission occurs along the axis of the active medium.

This in turn is amplified by the mirrors at either end of the active medium to amplify the laser beam. Among the two mirrors of the resonator, one is totally reflecting and the other is partially reflecting.

2.10 Semiconductor laser

There are two types of semiconductor lasers
   a). Homojunction semiconductor laser
   b). Heterojunction semiconductor laser

2.10.1 Homo-junction semiconductor laser
Characteristics
   a). **Type:** Solid state semiconductor laser.
   b). **Active medium:** pn-junction diode.
   c). **Pumping method:** Direct conversion method
   d). **Power output:** 1 mW.
   e). **Nature of output:** continuous wave or pulsed output.
   f). **Wavelength of Output:** 8.3 μm to 8.5 μm (infrared)

Principle

When the P-N junction diode is forward biased, the electrons from N-region and holes from P-region cross the junction and then recombine with each other to emit photon. The photon emitted during recombination stimulates other electrons and holes to recombine. As a result, laser beam is produced.

Construction

The active medium of the homojunction semiconductor laser is a P-N junction diode made from the single crystal of GaAs. The crystal GaAs is taken in the form of a platelet and consists of two regions n-type and p-type. The metal electrodes are fixed at the top and bottom of the platelet. The end faces of the P-N junction are well polished and parallel to each other. They act as an optical resonator. The homojunction semiconductor laser is shown in Fig. 2.13.

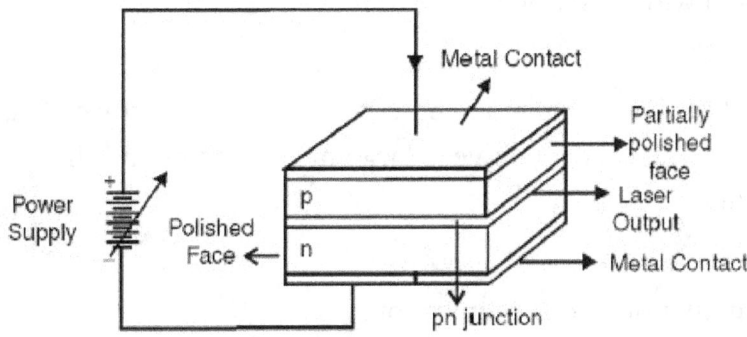

Fig. 2.13 Homojunction semiconductor laser

Working

When the P-N junction is forward biased, the electrons and holes are injected into the junction region. The region around the junction contains large amount of electrons and holes within the conduction band and valance band. Hence the population inversion is achieved.

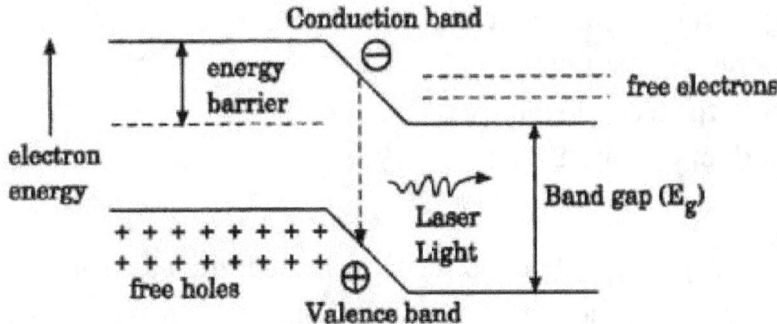

Fig. 2.14 Energy level digram.

As shown in Fig. 2.14, the recombination of electrons and holes takes place. During recombination, light photons are produced. The emitted photons produce a chain of stimulated emission to produce more intense light photons.

After gaining enough strength, laser beam of wavelength 8.4 μm is emitted from the junction. The wavelength of laser light is given by $\lambda = hc/E_g$

Advantages
- a). The arrangement is simple and compact.
- b). It exhibits high efficiency.
- c). The laser output can be easily increased by controlling the junction current
- d). It is operated with lesser power than ruby and CO_2 laser.

Disadvantages
- a). It is difficult to control the mode pattern and mode structure of laser.
- b). Threshold current density is very large and it has poor coherence and poor stability.

Applications
- a). It is used in fiber optic communication.
- b). It is used to heal the wounds by infrared radiation
- c). It is used as a pain killer
- d). It is used in laser printers and CD writing and reading

2.10.2 Hetero-junction semiconductor laser

Characteristics
- **a). Type**: Solid state semiconductor laser.
- **b). Active medium**: pn-junction diode.
- **c). Pumping method**: Direct conversion method
- **d). Power output:** 1 mW.
- **e). Nature of output:** continuous wave.
- **f). Wavelength of Output:** 8.0 μm

Principle

When the P-N junction diode is forward biased, the electrons from N-region and holes from P-region cross the junction and then recombine with each other to emit photon. The photon emitted during recombination stimulates other electrons and holes to recombine. As a result, laser beam is produced.

Construction

Hetero-junction semiconductor laser is shown in Fig. 2.15. It consists of five layers. A layer of GaAs P-type (3rd layer) acts as active region. This layer is placed between GaAlAs P-type (2nd layer) and GaAlAs N-type (4th layer). The metal electrodes are fixed at the top and bottom of the layer. The end faces of the P-N junction of 1st layer (GaAs P-type) and 5th layer (GaAs N-type) are well polished and parallel to each other. They act as an optical resonator.

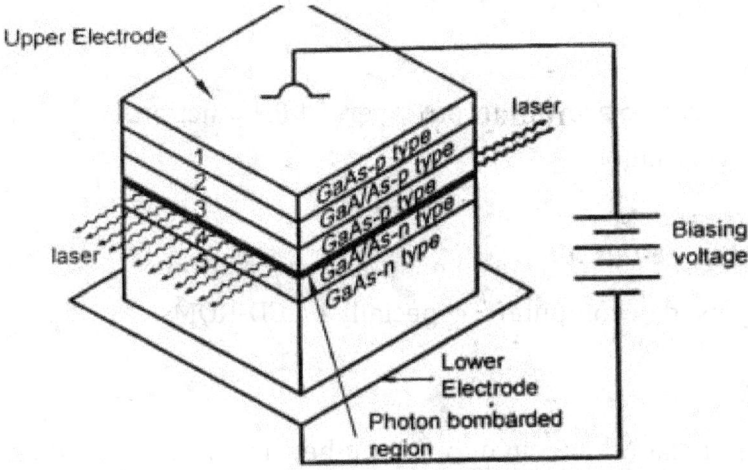

Fig. 2.15 Heterojunction laser

Working

When the P-N junction is forward biased, the electrons and holes are injected into the junction region. The region around the junction contains large amount of electrons and holes within the conduction band and valance band. Hence the population inversion is achieved.

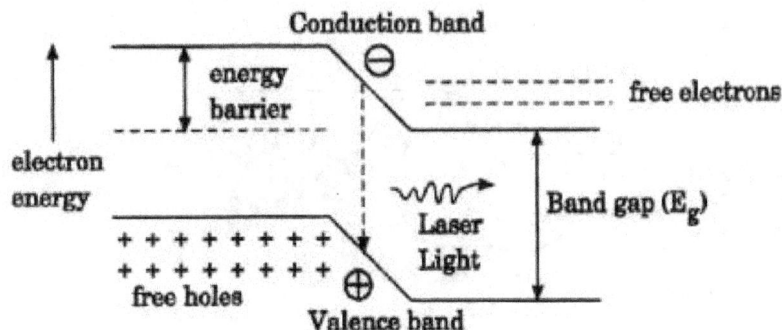

Fig. 2.16 Energy level digram

As shown in Fig. 2.16, the recombination of electrons and holes takes place. During recombination, light photons are produced. The emitted photons produce a chain of stimulated emission to produce more intense light photons.

After gaining enough strength, laser beam of wavelength 8.0 μm is emitted from the junction. The wavelength of laser light is given by λ = hc/E$_g$

Advantages
    a). It produces continuous wave output.
    b). The power output is very high.

Disadvantages
    a). It is very difficult to grow different layers of PN junction.
    b). The cost is very high.

Applications
    a). This type of laser is mostly used in optical applications
    b). It is widely used in computers, especially on CD-ROMs.

2.11 Fibre optics

The communication through optical fiber is known as fiber optics communication.

2.11.1 Structure of optical fiber

Optical fiber consists of an inner cylinder made of glass or plastic called core. The core has high refractive index (η_1). It is surrounded by a cylindrical shell of glass or plastic called cladding. The cladding has low refractive index (η_2). This cladding is covered by a jacket which protects fiber from moisture and abrasion. The structure of optical fiber is shown in Fig. 2.17.

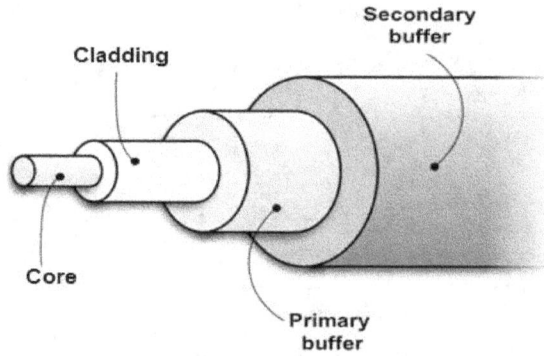

Fig. 2.17 Structure of optical fiber

Waves and Fiber Optics

Principle

The light is transmitted through an optical fiber by the principle of total internal reflection.

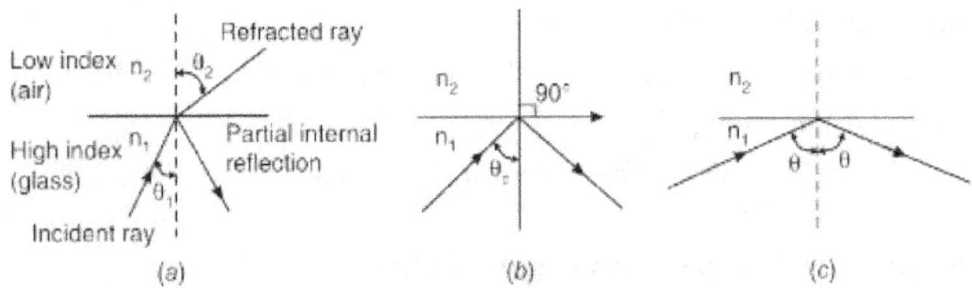

Fig. 2.18 Light propagation into optical fiber: (a) $\theta_r > \theta_i$, (b) $\theta_i = \theta_c$, (c) $\theta_i > \theta_c$

Case (i)

When an angle of refraction (θ_r) is greater than the angle of incidence (θ_i), i.e. $\theta_r > \theta_i$, the light ray is refracted into the rarer medium of refractive index (η_2). It is shown in Fig. 2.18(a).

Case (ii)

When an angle of incidence (θ_i) is equal to critical angle (θ_c), i.e. $\theta_i = \theta_c$ and $\theta_r = 90°$, then the light ray is refracted at the interface and it just emerges along the boundary. It is shown in Fig. 2.18(b)

Case (iii)

When $\theta_i > \theta_c$, then the light ray is reflected back into the same medium by total internal reflection. It is shown in Fig. 2.18(c).

2.11.2 Expression for critical angle (θ_c)

For the refraction of light, a relation between the angle of incidence (θ_i) and angle of refraction (θ_r) is given by Snell's law.

$$\eta_1 \sin\theta_i = \eta_2 \sin\theta_r \quad (1)$$

For total internal reflection $\quad \theta_i = \theta_c$ and $\theta_r = 90° \quad (2)$

Substituting equation (2) in equation (1), $\eta_1 \sin\theta_c = \eta_2 \sin 90°$

$$\sin\theta_c = \frac{\eta_2}{\eta_1} \sin 90°$$

$$\sin\theta_c = \frac{\eta_2}{\eta_1} \qquad \therefore \sin 90° = 1$$

$$\theta_c = \sin^{-1}\left(\frac{\eta_2}{\eta_1}\right) \qquad (3)$$

Equation (3) represents the expression for critical angle.

Conditions for total internal reflection

a). The refractive index (η_1) of core should be greater than the refractive index (η_2) of cladding.

b). The light ray should be incident at an angle greater than critical angle

2.12 Principle of light propagation in optical fibers

Consider the light propagation in an optical fiber as shown in the Fig. 2.19. An incident ray (AO) enters into a core at an angle (θ_0) to fiber axis. The incident ray is refracted along OB at an angle of refraction (θ_r) in core. The refracted ray falls on the interface of core and cladding at the critical angle of incidence $(\theta_c = 90° \quad \theta_r)$ and it moves along BC.

The light ray enters at angle of incidence greater than θ_0 at O, incident at B at an angle less than the critical angle. This light ray is refracted into the cladding region and it is absorbed.

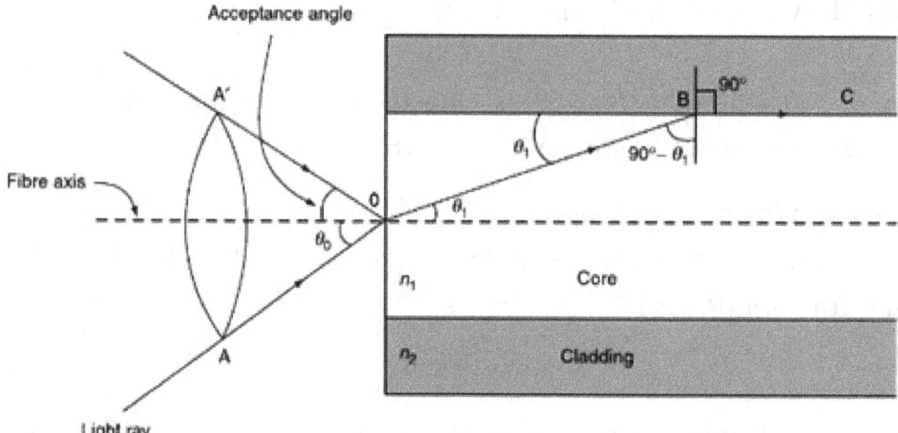

Fig. 2.19 Principle of light propagation in optical fiber

Let η_1, η_2 and η_0 be the refractive indices of the core, cladding and surroundings respectively.

Applying Snell's law of refraction at O,

$$\eta_0 \sin\theta_0 = \eta_1 \sin\theta_r \qquad (1)$$

$$\sin\theta_0 = \frac{\eta_1}{\eta_0}\sin\theta_r$$

$$\sin\theta_0 = \frac{\eta_1}{\eta_0}\sqrt{1 - \cos^2\theta_r} \qquad (2)$$

At the point B on the interference of core and cladding, angle of incidence

$$\theta_c = 90° - \theta_r$$

Applying Snell's law of refraction at B,

$$\eta_1 \sin(90° - \theta_r) = \eta_2 \sin 90°$$

$$\eta_1 \cos\theta_r = \eta_2$$

i.e. $\cos\theta_r = \sin(90° - \theta_r)$ and $\sin 90° = 1$

$$\cos\theta_r = \frac{\eta_2}{\eta_1} \qquad (3)$$

substituting equation (3) in equation (2), we get

$$\sin\theta_0 = \frac{\eta_1}{\eta_0}\sqrt{1 - \frac{\eta_2^2}{\eta_1^2}}$$

$$\sin\theta_0 = \frac{\sqrt{\eta_1^2 - \eta_2^2}}{\eta_0} \qquad (4)$$

$$\theta_0 = \sin^{-1}\frac{\sqrt{\eta_1^2 - \eta_2^2}}{\eta_0} \qquad (5)$$

Equation (5) represents the expression for acceptance angle. When the medium surrounding the fiber is air, η_0 equals to 1.

∴ Equation (5) can be written as $\theta_0 = \sin^{-1}\sqrt{\eta_1^2 - \eta_2^2}$ \qquad (6)

Acceptance angle can be defined as the maximum angle (θ_0) at which a ray of light can enter through one end of the fiber and still be totally internally reflected.

2.13 Numerical aperture (NA)

The sine of the acceptance angle of the fiber is called as numerical aperture.

It is given by $\qquad NA = \sin\theta_0 \qquad$ (7)

Comparing equations (4) and (7), we get

$$NA = \frac{\sqrt{\eta_1^2 - \eta_2^2}}{\eta_0} \qquad (8)$$

If $\eta_0 = 1$ for air medium,

$$NA = \sqrt{\eta_1^2 - \eta_2^2} \qquad (9)$$

Equation (9) represents the expression for numerical aperture.

2.14 Types of optical fibers

Optical fibres are classified based on the following categories
1. Material
2. Number of modes
3. Refractive index

1. Classification based on material

Based on material, optical fibers are classified into two types. They are

a). **Glass fibres:** They are made up of a glasses and metal oxides.
   Examples: (i) SiO_2-GeO_2 (core) and SiO_2 (cladding),
   (ii) SiO_2 (core) and P_2O_3-SiO_2 (cladding)

b). **Plastic fibres:** They are made up of plastics.
   Examples: (i) Polystyrene (core) and methyl methacrylate (cladding)
   (ii) PMMA (core) and co-polymer (cladding)

2. Classification based on number of modes

Based on number of modes, optical fibers are classified into two types. They are (a) Single mode fiber and (b) Multimode fiber.

a). **Single mode fiber:**

It is shown in the Fig. 2.20 and has the following characteristics.
   (i). It allows only one mode of light propagation
   (ii). Radius of core is small

(iii). It carries information to longer distance

(iv). Intermodal dispersion is free in single mode fiber

(v). Coupling process of this fiber is not easy

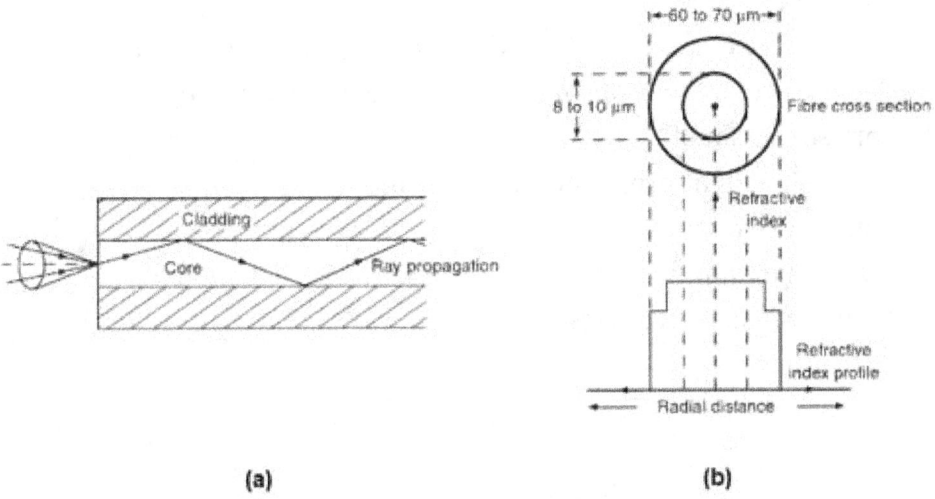

Fig. 2.20 (a) Single mode fiber and (b) cross-sectional view

b). Multimode fiber:

It is shown in the Fig. 2.21 and has the following characteristics.

(i). It allows large number of modes of light propagation

(ii). Radius of core is large

(iii). It carries information to shorter distance

(iv). Intermodal dispersion exists in multimode fiber

(v). Coupling process of this fiber is easy

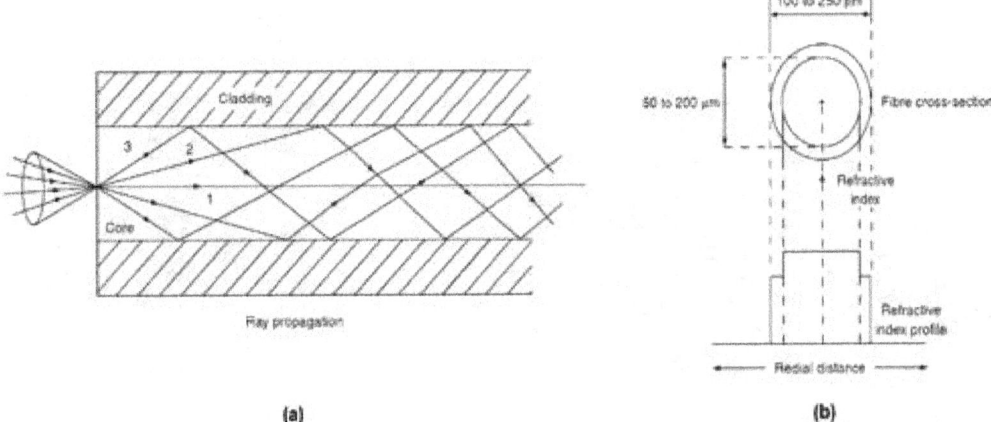

Fig. 2.21 (a) Multimode optical fiber and (b) cross-sectional view

3. Classification based on refractive index

Based on refractive index, optical fibers are classified into two types. They are (a) step index fiber and (b) Graded index fiber.

(a) Step index fiber:

It is shown in the Fig. 2.22 and has the following characteristics.

(i). The refractive index of core and cladding varies step by step

(ii). Core size is small

(iii). The path of light propagation is zig-zag

(iv). It has more distortion

(v). It has lower bandwidth

(vi). Numerical aperture is less

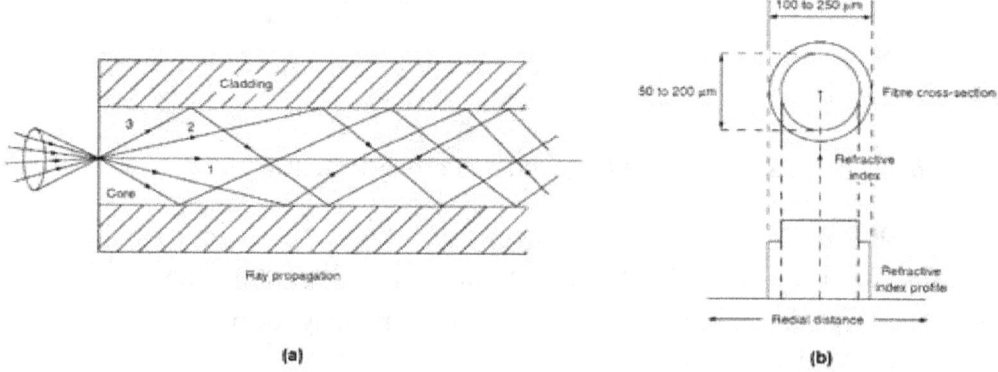

Fig. 2.22 (a) step index optical fiber and (b) cross-sectional view

(b) Graded index fiber

It is shown in the Fig. 2.23.

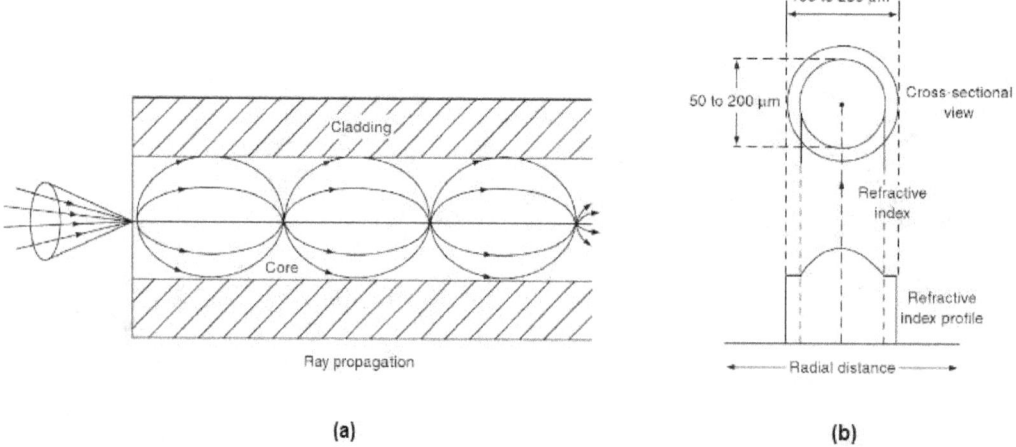

Fig. 2.23 (a) graded index optical fiber and (b) cross-sectional view

Waves and Fiber Optics

The graded index optical fiber has the following characteristics.
   (i). The refractive index of the core is maximum along the fibre axis and it gradually decreases towards core-cladding interface.
   (ii). Core size is large
   (iii). The path of light propagation is helical
   (iv). It has less distortion
   (v). It has higher bandwidth
   (vi). Numerical aperture is high

2.15 Losses associated with optical fiber

When light signal propagates through a fiber, it can be suffered by loss of amplitude and change in shape. Losses in optical fibre can be divided into two categories: A). attenuation and B). dispersion.

A). Attenuation:

The loss of amplitude of light signals is known as attenuation. It is further classified into two types: (a) intrinsic attenuation and (b) extrinsic attenuation.

(a) Intrinsic attenuation:

When a light ray falls on impurity present in the fibre, it may be absorbed or scattered. This is known as intrinsic attenuation. It is further divided into two categories: (i) Material absorption and (ii) Rayleigh scattering.

(i) Material absorption: It is due to imperfection of atomic structure and impurities present in the optical fiber.

(ii) Rayleigh scattering: It is due to local microscopic density variations in the optical fiber. It causes the variations of refractive index. They act as obstructions and scatter light in all directions.

(b) Extrinsic attenuation:

The bending effect affects the refractive index and the critical angle of the light, So total internal reflection is not satisfied. It is known as extrinsic attenuation. It is further divided into two categories: (i) macro-bend and (ii) micro-bend.

(i) Macro-bend: It is due to the light radiation does not satisfy the condition for the total internal reflection at the corner. It is also known large scale bend.

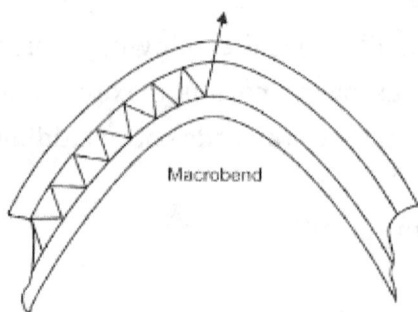

Fig. 2.24 Macro bend

(ii) Micro-bend: It is due to non-uniform pressures created during the manufacturing. It is also known as small scale bend.

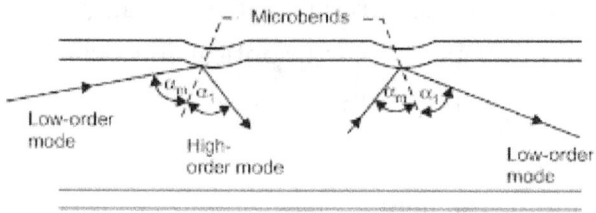

Fig. 2.25 Micro-bend

B). Dispersion: The change in shape of light signals is known as dispersion. It is further classified into three categories: (a) intermodal dispersion, (b) material dispersion and (c) waveguide dispersion

(a) Intermodal dispersion: It is shown in Fig. 2.26 and has the following characteristics.

(i). The lower modes travel a greater distance than the higher order modes

(ii). The path of lower mode is shorter while the path of higher mode is longer

(iii). Lower mode reaches the end of the fiber earlier while the higher mode reaches after some time delay.

(iv). It causes the signal distortion.

(b) Material dispersion: It is shown in Fig. 2.26 and has the following characteristics.

(i). A light is composed of a group of components of different wavelengths. The different wavelength components will propogate at different velocity along the fibre.

(ii). The shorter wavelength components travel slower than the long wavelength components

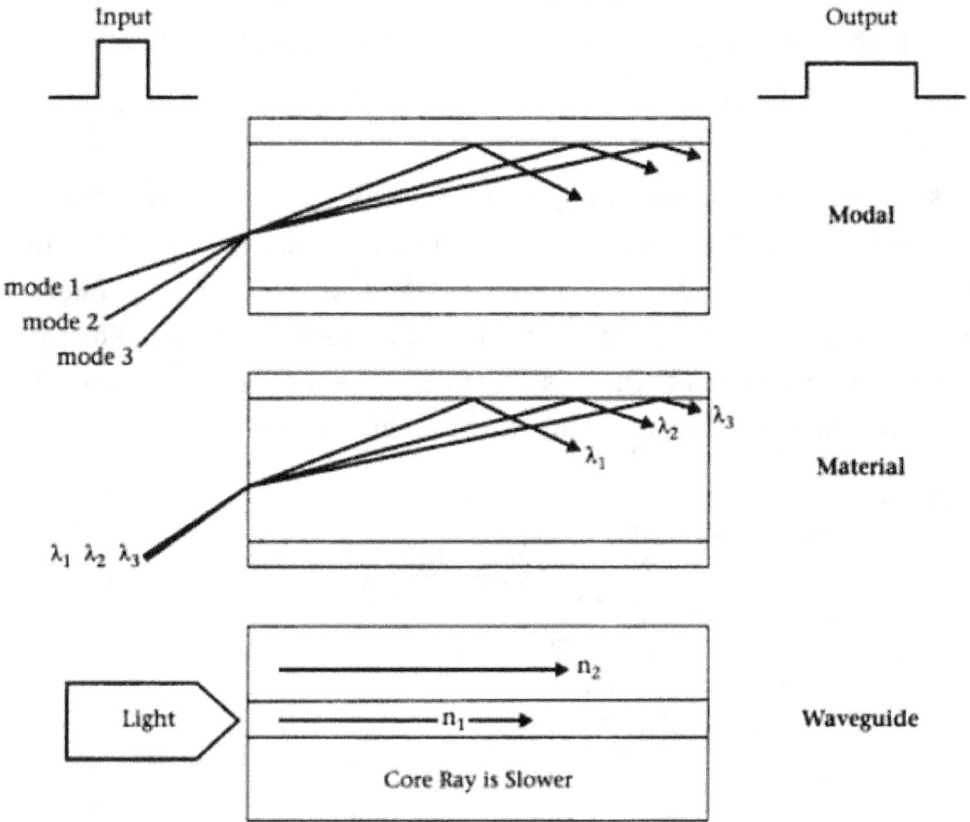

Fig. 2.26 various dispersions in optical fiber

(c) Waveguide dispersion: It is shown in Fig. 2.26 and has the following characteristics. It is due to the guiding property of the fiber and different angles of incidence at the core-cladding interface.

2.16 Fiber optic sensor

It is a transducer which converts one form of energy into another form. There are two types of sensors:

(i) Intrinsic (or) active sensor
(ii) Extrinsic (or) passive sensor.

(i) Intrinsic sensor

Physical parameters such as temperature, pressure, etc is directly sensed by optical fiber itself. Example: pressure sensor.

Pressure sensor
Principle

When an optical fiber is subjected to pressure variations, then the change in length and refractive index of optical fiber occurs. It causes the change in phase of light at end of fiber.

Construction

Pressure sensor is shown in Fig. 2.27. It consists of laser source which emits laser light beam. A beam splitter is inclined at an angle of 45° with respect to the direction of laser beam from the laser source. Two optical fibers, reference fiber and test fiber, are used in this system. Reference fiber is isolated from the environment whereas test fiber is kept in the environment. Lens systems are used to split and collect the beam of laser light.

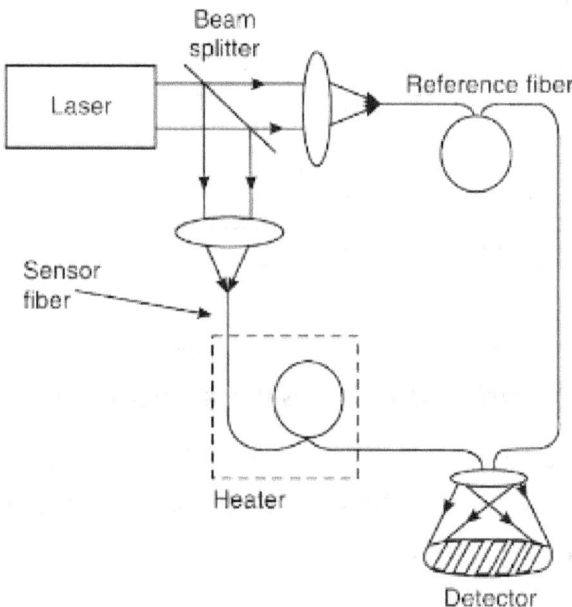

Fig. 2.27 Pressure sensor

Working

A monochromatic light beam is emitted from the laser and made to fall on the beam splitter which divides laser beam into two beams called reference beam and test beam. The beams after passing through the reference fiber and test fiber are made to fall on the lens (L2) which produces the path difference. The path difference causes the inference as shown in Fig. 2.27. Thus the change in pressure or temperature can be studied with the help of the interference pattern.

(ii) Extrinsic sensor

Separate sensing element will be used and the fiber will act as a guiding media to the sensors. Example: displacement sensor.

Displacement sensor
Principle

Light is sent through a transmitting fiber and is made to fall on a moving target. The reflected beam from the target is sensed by a detector with respect to the intensity of light reflected from the target and hence the displacement of the target is measured.

Construction:

Displacement sensor is shown in Fig. 2.28. It consists of a bundle of transmitting fibres connected to a laser source and a bundle of receiving fibres connected to a detector.

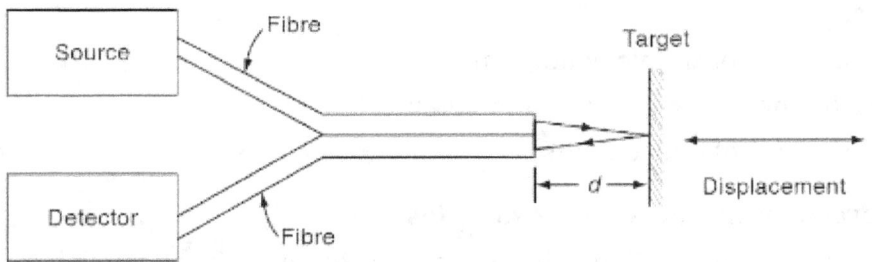

Fig. 2.28 Displacement sensor

Working

The light beam from the laser source is transmitted through transmitting fibre and it falls on the moving target. The reflected beam from the target is made to pass through the receiving fibre and this light is detected by detector. The intensity of the light received back depends on the displacement of the target.

If the received intensity increases, it denotes that the target is mving towards the sensor. If the intensity of light received decreases, it denotes that the target is moving away from the sensor. Thus the displacement of the target is measured from the intensity of the light received.

PART-A TWO MARKS

1. Define free oscillations. Give examples.

When a body vibrates with its natural frequency, it executes free oscillations.

Examples:
  a) Oscillations of simple pendulum
  b) Vibrations of tuning fork
  c) Vibrations in stretched string

2. Define damped oscillations. Give examples.

When an oscillation occurs, damping force may arise due to friction of air resistance offered by the medium. The amplitude of oscillations decreases with time and finally becomes zero. Such oscillations are called damped oscillations.

Examples:
  a) Oscillations of simple pendulum
  b) Oscillations of a coil in galvanometer
  c) Oscillations of tank circuits

3. Define forced oscillations. Give examples.

When a body is oscillated by a periodic force of frequency than its natural frequency, the oscillations are called forced oscillations

Example: sound boards of stringed instruments.

4. Define plane progressive waves

Progressive wave is originating from a point source and propagating through an isotrophic medium travel in all directions with same velocity.

5. Name the different processes occur when light radiation interacts with a material.
  a). Stimulated absorption
  b). Spontaneous emission
  c). Stimulated emission.

6. State Einstein's coefficients.

The proportionality constants A_{21}, B_{12} and B_{21} for upward and downward transitions of light photon are known as Einstein's Coefficients A and B

7. What is meant by population inversion?

The state of achieving more number of atoms in the excited state (N_2) than that of the atoms in the ground state (N_1) is called population inversion. i.e. $N_2 > N_1$

8. What are different methods of achieving population inversion?
a). Optical pumping
b). Electric discharge
c). Inelastic atom-atom collision
d). Direct Conversion
e). Chemical process

9. Define resonant cavity.

It is a cavity in which an active material is kept between two mirrors. One of the mirrors is made partially reflecting light beam while the other mirror is highly reflecting one.

10. What are the conditions required for the laser?
a). Population inversion should be achieved
b). Stimulated emission should be predominant over spontaneous emission.

11. Distinguish between homojunction and heterojunction semiconductor lasers.

Sl. No	Homojunction semiconductor laser	Heterojunction semiconductor laser
1.	It is made by single crystalline	It is made by different crystalline materials
2.	Power output is low	Power output is high
3.	Pulsed output	Continuous output
4.	Cost is less	Cost is more
5.	Life time is less	Life time is more

12. Explain stimulated emission.

It is a process in which there is an emission of two photons whenever an atom is transferred from a higher energy state to a lower energy under the influence of an external force.

13. **Name any two uses of lasers in medicine.**
    a). Laser is used for the treatment of detached retina.
    b). It is used to make precise cut in bones.
    c). It is used to disintegrate urinary stones
    d). It is used to therapeutic applications

14. **Name the properties of laser, which are making it suitable for industrial applications.**
    a). The laser beam is highly directional.
    b). It has high intensity.
    c). It has purely monochromatic.
    d). It has coherence.

15. **Define the following: (i) Numerical aperture and (ii) Acceptance angle.**
    (i). Numerical aperture: The sine of acceptance angle of the fiber is known as numerical aperture.
    (ii). Acceptance angle: The maximum angle at which a light ray is still totally internally reflected into an optical fiber

16. **Define attenuation in an optical fiber and mention its unit.**
    The loss of optical power as light travels through a fiber is known as attenuation. It is defined as the ratio of the optical power output (Pout) from a fiber of length 'L' to the power input (Pin). Its unit is dB/km.

17. **What are the uses of optical fibers.**
    Optical fibers are
    a). used in the long haul communication.
    b). used in laparoscope and endoscope
    c). used as sensors to measure the temperature and pressure.

18. **Name the principle behind the transmission of light wave through optical fiber (or) Define total internal reflection.**
    When angle of incidence is further increased above the critical angle, the ray is reflected back into the core. This phenomenon is called as Total internal reflection.

19. **State the principle behind the optical fiber sensor.**

    The fibre optic sensor is a transducer which converts one form of energy (pressure, temperature, etc) into another.

20. **Name the losses occurs during optical fiber communication.**

    (a) Attenuation and (b) dispersion.

PART- B

1. Obtain the differential equation of damped harmonic oscillation and discuss the special cases of oscillatory motion.
2. Discuss the theory of forced harmonic oscillations. How does sharpness of resonance depend on damping?
3. Derive the expression for the wave equation of a plane progressive wave.
4. Derive an expression for Einstein's coefficient of spontaneous and stimulated emissions.
5. Explain the principle, construction and working of a homo-junction and hetero-junction semiconductor diode lasers. Mention their advantages and disadvantages.
6. Define numerical aperture and derive an expression for numerical aperture and angle of acceptance of fiber in terms of refractive index of the core and cladding.
7. How optical fibers are classified based on modes, material and refractive index profile?. Explain them in detail.
8. Discuss the losses associated with the optical fibers.
9. Explain the construction and working of pressure sensor.
10. Explain the construction and working of displacement sensor.

UNIT-III
THERMAL PHYSICS

3.1 Transfer of heat energy
3.2 Thermal expansion of solids
Types of expansions
(i) Linear expansion (expansion in length)

The coefficient of linear expansion of a solid is increase in length produced per unit length of the solid when its temperature is raised by 1 K. It is denoted by α.

$$\alpha = \frac{L_2 - L_1}{L_1(T_2 - T_1)}$$

(ii) Superficial expansion-Expansion in area

The coefficient of superficial expansion of a solid is increase in area produced per unit area of the solid when its temperature is raised by 1 K. It is denoted by β.

$$\beta = \frac{A_2 - A_1}{A_1(T_2 - T_1)}$$

(iii) Cubical expansion-Expansion in volume

The coefficient of cubical expansion of a solid is increase in volume produced per unit volume of the solid when its temperature is raised by 1 K. It is denoted by υ.

$$\upsilon = \frac{V_2 - V_1}{V_1(T_2 - T_1)}$$

Applications of expansion of solids
a) A gap is left at the joint of two rails
b) The most applications are expansion joints and bio-metallic strips
c) The gaps are provided in concrete highways and bridges for temperature variations

Thermal Physics

3.3 Thermal expansion of liquids

(i) Real or absolute expansion

The coefficient of real or absolute expansion of a liquid is real increase in volume produced per unit volume of the liquid when its temperature is raised by 1 K. It is denoted by υ_r

$$\upsilon_r = \frac{V_2 - V_1}{V_1(T_2 - T_1)}$$

(ii) Apparent expansion

The coefficient of apparent expansion of a liquid is observed increase in volume produced per unit volume of the liquid when its temperature is raised by 1 K. It is denoted by υ_a

$$\upsilon_a = \frac{V_2 - V_1}{V_1(T_2 - T_1)}$$

3.4 Expansion joints

It is an assembly designed to absorb the heat induced expansion or contraction of a pipeline, duct or vessel.

Types of expansion joints

(i) Metallic expansion joints

These expansion joints are provided for thermal expansion and relative movement in pipelines, containers and machines. It is shown in Fig. 3.1.

Fig. 3.1 Metallic expansion joints

The metallic expansion joints are classified into three types. They are

    a). Axial expansion joints

    b). Angular expansion joints

    c). Lateral expansion joints

(ii) Wall expansion joints

They are provided in large walls to allow for the expansion of concrete due to temperature changes. It is shown in Fig. 3.2.

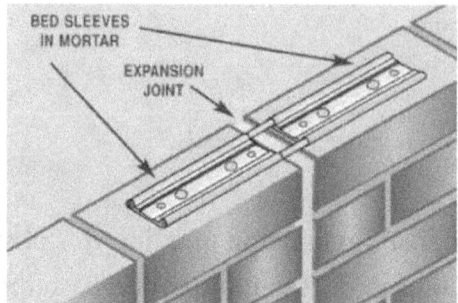

Fig. 3.2 Wall expansion joints

3.5 Bimetallic strips

A bimetallic strip is known as strip made of two metals of different expansion coefficients joined together.

Principle

When the strip is heated, the metals expand at different rates, causing the strip to bend

Description

The bimetallic strip is shown in Fig. 3.3. It consists of two strips of different metals which expand at different rates as they are heated, usually iron and copper or iron and brass. Two strips of different metals with equal lengths are joined together to form a bimetallic strip.

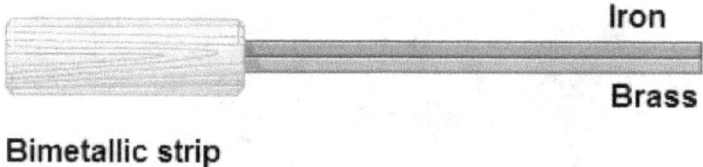

Fig. 3.3 Bimetallic strips

Description

The illustration of bimetallic strip is shown in Fig. 3.4. When the bimetallic strip, made by brass and iron, is heated by a burner, the iron strip bends to form a curve but the brass strip remains on the outside of the curve. The brass strip is longer than the iron strip after heating.

Thermal Physics

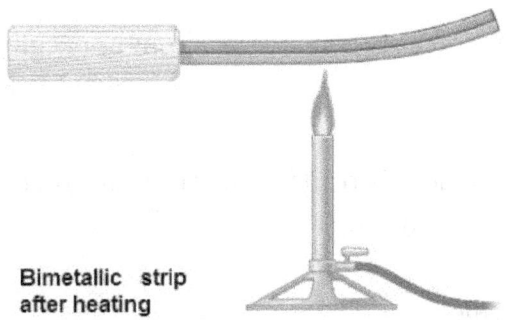

Bimetallic strip after heating

Fig. 3.4 Expansion of bimetallic joints

Applications

Bimetallic strips are used as
   a). bimetallic thermo-switches
   b). bimetallic thermostats
   c). bimetallic thermometer
   d). bimetallic sensors

3.6 Three modes of transmission of heat

Heat transfer is known as that the transmission of heat takes place from a hot surface to a cold surface. As shown in Fig. 3.5, there are three modes of transmission of heat. They are (i) conduction, (ii) convection and (iii) radiation.

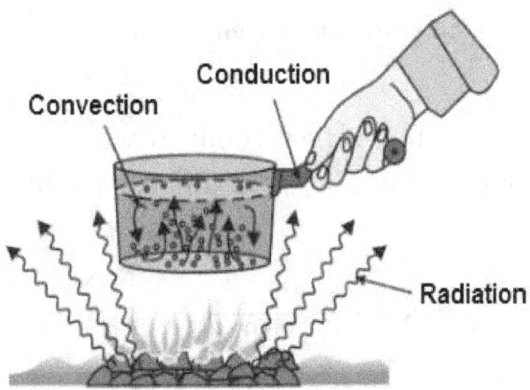

Fig. 3.5 Three modes of heat transfer

(i) Conduction

It is known as that the heat is transferred from hot region to cold region through the substance without the motion of particles.

Thermal Physics

(ii) Convection
It is known as that the heat is transferred from hot region to cold region by the motion of particles.

(iii) Radiation
It is known as that the heat is transferred from hot region to cold region directly without any material medium.

3.7 Heat conduction in solids
The heat is transmitted from a body of higher temperature to that of lower temperature.

Experiment
When a metal rod is heated at one end, heat gradually flows along the length of the rod and other end of the rod also becomes hot after some time. It shows that heat has travelled through the molecules of the rod from one end to other end. The molecules in the rod remain fixed in their mean position.

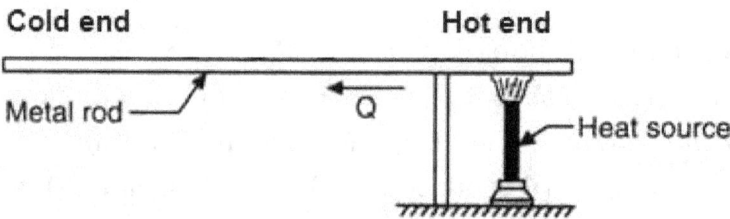

Fig. 3.6 Heat conduction in solids

3.8 Thermal conductivity
It is defined as the amount of heat conducted per second through unit area of the material when unit temperature gradient is maintained. Unit is $Wm^{-1}K^{-1}$. It is represented by K.

$$K = \frac{Qx}{A(\theta_2 - \theta_1)t}$$

3.9 Forbes method (Theory and experiment)
It is the earliest method to find the absolute thermal conductivity of metals.

Theory
Consider a long rod which is heated at one end and a steady state is reached after some time. The rod has a series of holes into which

thermometers are fitted. These thermometers record temperatures at different points along the rod as shown in Fig. 3.7. When the steady state is reached, the temperature (θ) shown by the thermometers of the rod and their respective distances (x) from the hot end are noted.

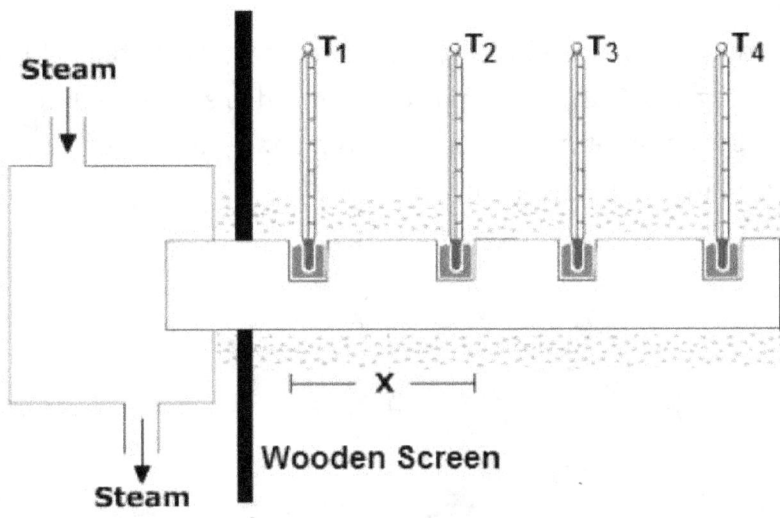

Fig. 3.7 Forbe's method

A graph is drawn with temperature along the Y-axis and the distance of the thermometers from the steam chamber along X-axis. As shown in Fig. 3.8, a curve is obtained. A tangent is drawn to the curve at a point corresponding to the point B near hot end. The temperature gradient at B,

$$\frac{d\theta}{dx} = \frac{AB}{BC} \tag{1}$$

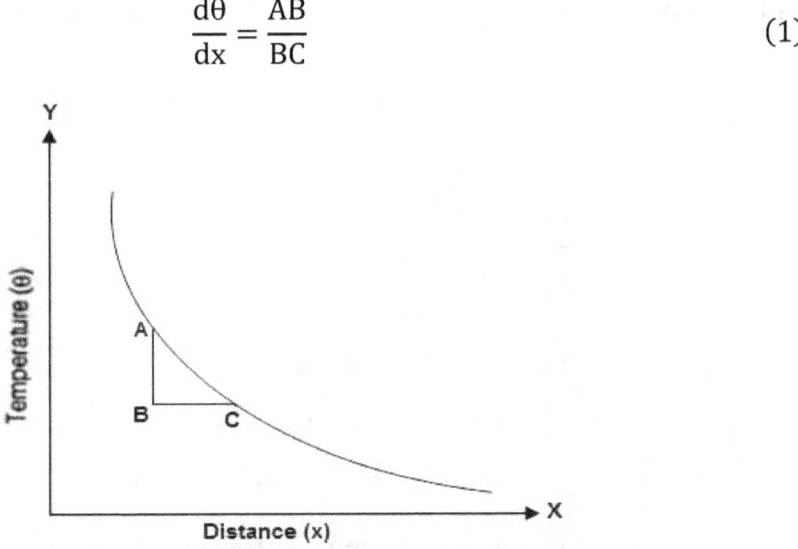

Fig. 3.8 a graph of temperature (θ) vs distance (x)

When steady state is reached, the heat conducted,

$$Q_1 = KA\frac{d\theta}{dx} \quad (2)$$

Where A is area of cross-section of the metal rod and K is the coefficient of thermal conductivity.

A piece of the rod is heated to the same temperature as that of the hot end. It is shown in Fig. 3.9. The heated piece of the rod is suspended in air and allowed to cool. Its temperature is noted at regular intervals of time by a thermometer placed in a hole at the centre.

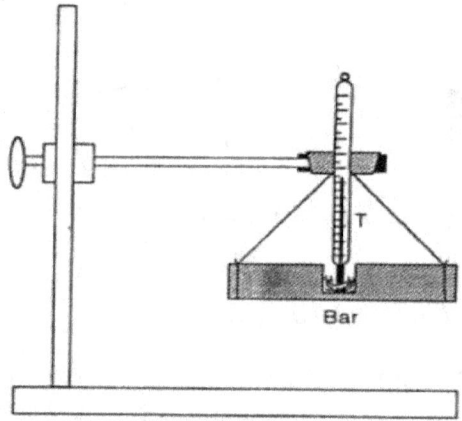

Fig. 3.9 A piece of rod is heated

A graph is drawn between temperature (θ) and time (t) as shown in Fig. 3.10. From this graph, the value of $\frac{d\theta}{dt}$ for various values of θ is determined.

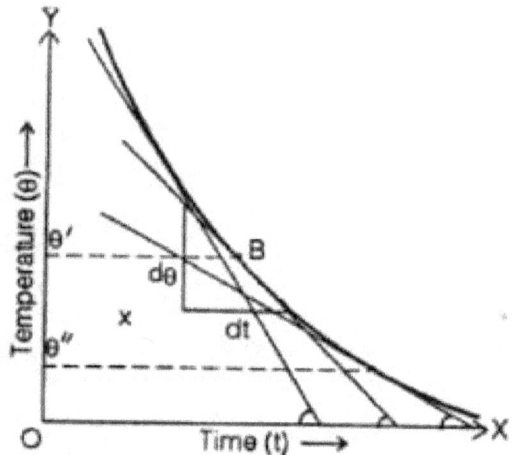

Fig. 3.10 A graph of temperature (θ) vs time (t)

Another graph is drawn with distance of thermometers from the hot end along X axis and the corresponding rate of cooling $\frac{d\theta}{dt}$ along Y axis as shown in Fig. 3.11.

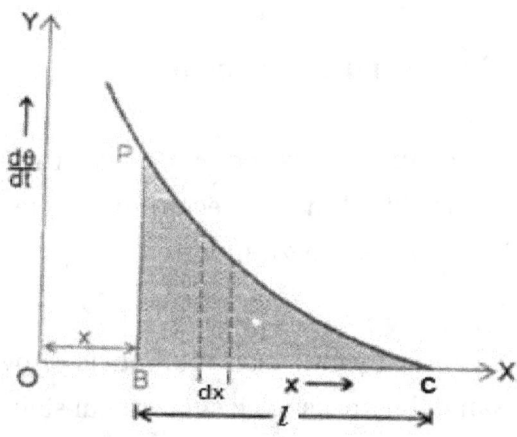

Fig. 3.11 a graph of cooling rate $\frac{d\theta}{dt}$ vs distance (x) of thermometers

The shaded area represents the value

$$= \int \frac{d\theta}{dt} \delta x \qquad (3)$$

If δx is the small area of the rod, ρ is the density of the metal, A is area of cross-section and $\rho A \delta x$ is the mass of the small area of the rod, then the total heat radiated per second, Q_2

$$= A\rho S \int \frac{d\theta}{dt} \delta x \qquad (4)$$

At equilibrium,

$$\begin{pmatrix} \text{Amount of heat conducted} \\ \text{per second} \\ \text{per unit cross section} \end{pmatrix} = \begin{pmatrix} \text{Amount of heat radiated} \\ \text{per second} \\ \text{per unit cross section} \end{pmatrix}$$

$$Q_1 = Q_2$$

$$KA \frac{d\theta}{dx} = A\rho S \int \frac{d\theta}{dt} \delta x$$

Thermal Physics

$$\therefore \quad K = \frac{\rho S \frac{d\theta}{dt}}{\frac{d\theta}{dx}} \qquad (5)$$

Equation (5) represents the thermal conductivity of material.

3.10 Lee's disc method (Theory and experiment)
Principle

The quantity of heat flowing across the bad conductor in one second is equal to the quantity of heat radiated in one second from the lower face area and edge area of the metal disc in the Lee's apparatus.

Construction

The experimental set up of Lee's disc method is shown in Fig. 3.12. The apparatus consists of same thickness of circular metal slab (Lee's disc) and a bad conductor in the form of the disc. They are placed one above the other. A hollow cylindrical steam chamber 'A' having the same diameter as that of the slab is placed over the bad conductor. Thermometers T_1 and T_2 are inserted into steam chamber and the slab to record temperature.

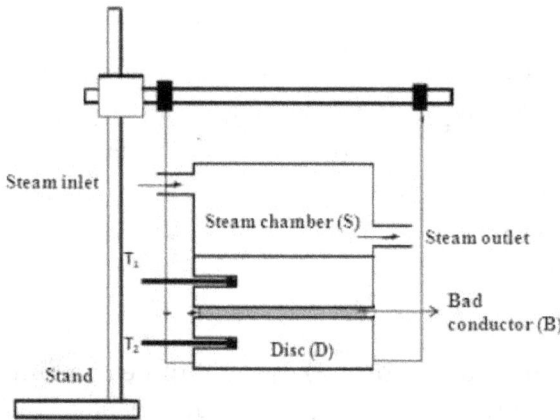

Fig. 3.12 The experimental set up of Lee's disc method

Working

When steam is passed through the steam chamber, heat is flowing from upper surface to lower surface of the bad conductor. Due to the poor thermal conductivity of bad conductor, the temperature of lower surface is less than the temperature of upper surface. When the steady state is reached, heat flowing across 'B' is taken up by 'D' and radiated away at the same rate from its lower face and its edges.

Observation and calculations

The heat flowing across the bad conductor 'B' per second is given by

$$Q = KA\left(\frac{\theta_1 - \theta_2}{x}\right) \quad (1)$$

where A is area of cross section of bad conductor, x is thickness of bad conductor, K is thermal conductivity, θ_1 and θ_2 are temperature at upper and lower surface of bad conductor.

The quantity of heat radiated per second from lower face area and edge area of metal disc is given by

$$Q' = (\pi r^2 + 2\pi rh)E \quad (2)$$

In steady state,

$$Q = Q'$$

$$KA\left(\frac{\theta_1 - \theta_2}{x}\right) = (\pi r^2 + 2\pi rh)E$$

$$K = \frac{(r+h)xE}{r(\theta_1 - \theta_2)} \quad (3)$$

When the temperature of the disc is 10°C above its steady state temperature, the steam chamber is removed and the disc is allowed to cool. A graph of temperature against time is plotted as shown in Fig. 3.13 and the slope of graph at the temperature at θ_1

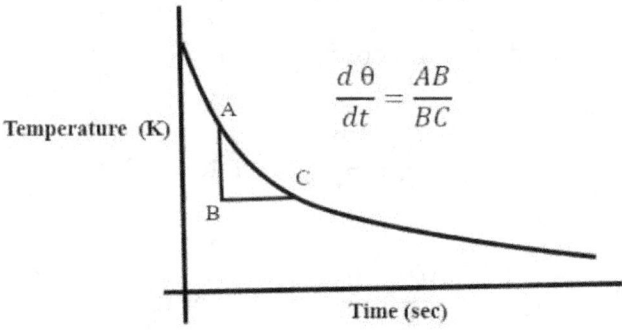

Fig. 3.13 a graph of temperature against time

If the rate of cooling is $\frac{d\theta}{dt}$, mass of the metallic disc is M and its specific heat is S, then the rate of loss of heat from its two faces and edge is $MS\dfrac{d\theta}{dt}$.

$$\therefore \text{ The emissivity (E)} = \dfrac{MS\dfrac{d\theta}{dt}}{\text{Area of surface}} = \dfrac{MS\dfrac{d\theta}{dt}}{2\pi r^2 + 2\pi rh} \qquad (4)$$

Substituting equation (4) in equation (3), we get
Thermal conductivity of bad conductor

$$K = \dfrac{MS\dfrac{d\theta}{dt}(r + 2h)}{2\pi r^2(\theta_1 - \theta_2)(r + h)} \qquad (5)$$

3.11 Conduction through compound media (series and parallel)
(i) Bodies in series

Let us consider a compound wall of two different materials 'A' and 'B' of thickness d_1 and d_2 as shown in Fig. 3.14. The temperature of end faces is θ_1 and θ_2 whereas the temperature of common surface is θ. Let K_1 and K_2 be the thermal conductivity of the materials A and B.

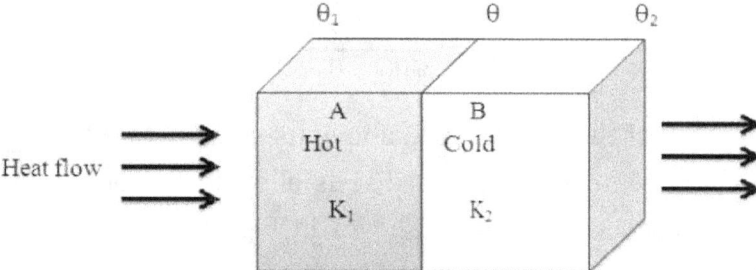

Fig. 3.14 Bodies in series

If Q is heat flowing per second through compound media, then

$$Q = \dfrac{K_1 A(\theta_1 - \theta)}{d_1} \text{ for material A} \qquad (1)$$

$$Q = \dfrac{K_2 A(\theta - \theta_2)}{d_2} \text{ for material B} \qquad (2)$$

When steady state is reached, equation (1) = equation (2)

$$\frac{K_1 A(\theta_1 - \theta)}{d_1} = \frac{K_2 A(\theta - \theta_2)}{d_2}$$

$$\frac{K_1 A \theta_1}{d_1} - \frac{K_1 A \theta}{d_1} = \frac{K_2 A \theta}{d_2} - \frac{K_1 A \theta_2}{d_2}$$

$$\frac{K_1 A \theta}{d_1} + \frac{K_2 A \theta}{d_2} = \frac{K_1 A \theta_1}{d_1} + \frac{K_1 A \theta_2}{d_2}$$

$$A\theta \left(\frac{K_1}{d_1} + \frac{K_2}{d_2}\right) = A\left(\frac{K_1 \theta_1}{d_1} + \frac{K_1 \theta_2}{d_2}\right)$$

$$\theta = \frac{\left(\frac{K_1 \theta_1}{d_1} + \frac{K_1 \theta_2}{d_2}\right)}{\left(\frac{K_1}{d_1} + \frac{K_2}{d_2}\right)}$$

(ii) Bodies in parallel

Let us consider a compound wall of two different materials 'A' and 'B' of thickness d_1 and d_2 as shown in Fig. 3.15. They are connected in parallel.

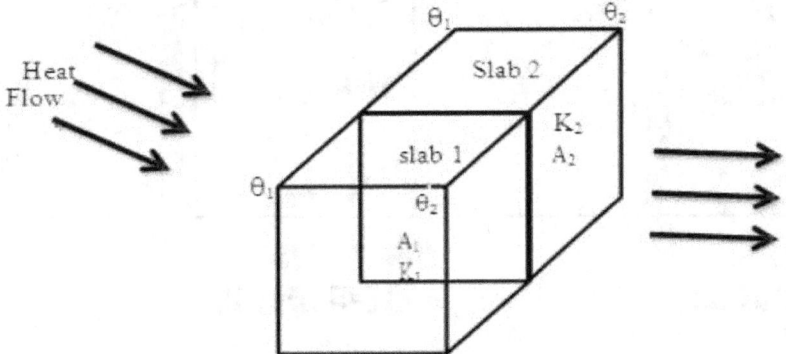

Fig. 3.15 Bodies in parallel

Let A_1 and A_2 be the areas of cross section of the materials. The total heat (Q) flowing through these materials per second will be the sum of Q_1 and Q_2

i.e. $Q = Q_1 + Q_2$

$$Q = \frac{K_1 A_1 (\theta_1 - \theta_2)}{d_1} + \frac{K_2 A_2 (\theta_1 - \theta_2)}{d_2}$$

Amount of heat flowing per second,

$$Q = (\theta_1 - \theta_2)\left(\frac{K_1 A_1}{d_1} + \frac{K_2 A_2}{d_2}\right)$$

The net amount of heat flowing per second through the bodies in parallel is given by

$$Q = (\theta_1 - \theta_2)\sum \frac{KA}{d}$$

3.12 Thermal insulation

Thermal insulation is used to reduce the flow of heat between outside and inside of a building. Due to thermal insulation, the energy consumption is also reduced to maintain artificial cooling in summer or artificial heating in winter in a room. It is illustrated in Fig. 3.16.

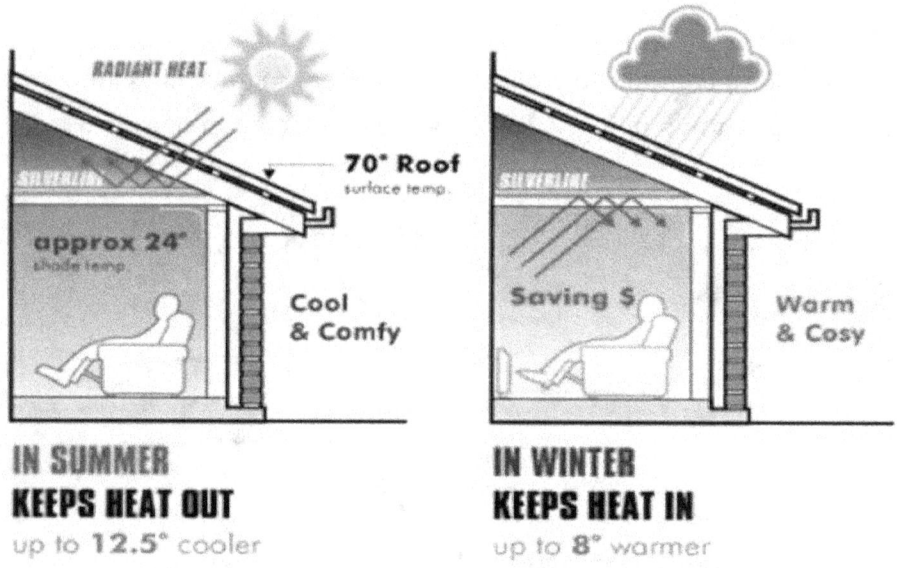

Fig. 3.16 Thermal insulation process

Thermal insulating materials

The materials which used to insulate thermally are known as thermal insulating materials. Based on the density, the thermal insulating materials are classified into two types

a). Less dense materials
b). High dense materials

Thermal Physics

Distinguish between less dense and high dense insulating materials

Sl. No	Less dense thermal insulating materials	High dense thermal insulating materials
1.	It has enormous resistance for conduction of heat	It has less resistance for conduction of heat
2.	It has low volumetric specific heat	It has high volumetric specific heat
3.	It does not absorb heat enormously	It absorbs heat enormously
4.	Example: Cork, softwood and carboard	Examples: Brick, glass, stone and concrete

Properties of thermal insulating materials

They should
- a). have low thermal conductivity
- b). absorb less moisture
- c). have adequate fire resistance
- d). have good stability
- e). have high specific heat
- f). be available at lower cost

Applications of thermal insulations

(i) Thermal insulation of exposed doors and windows

a). The insulating glass or double glass with air space may be provided for glass doors and windows. It will reduce heat transmission through doors and windows.

b). The protection in the form of sun breakers, weathersheds and projections curtains may be provided on the exposed doors and windows to reduce incidence of solar heat.

c). The flat roofs may be kept to cool by water which may be reduced the surface temperature.

d). The thermal insulation of flat roof may be provided by putting a layer.

(ii) Thermal insulation of exposed walls

A model illustration of thermal insulation of exposed walls is shown in Fig. 3.17.

a). The suitable thickness of wall may be provided.
b). The hollow wall or cavity wall construction may be adopted.
c). For partitions, an air space may be created by fixing hardboards.

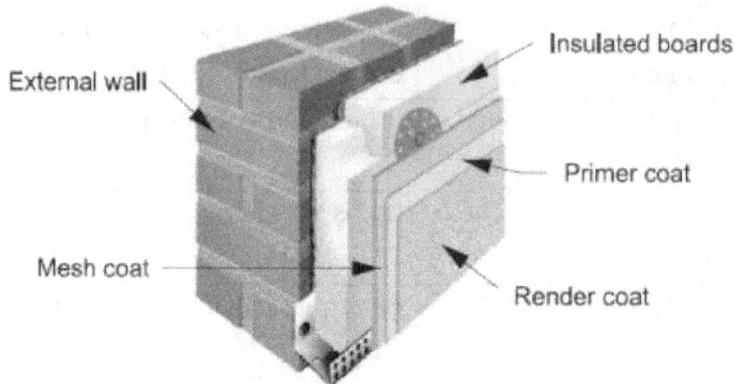

Fig. 3.17 Thermal insulation of exposed wall

(iii) Thermal insulation of exposed roofs

A model illustration of thermal insulation of exposed walls is shown in Fig. 3.18.

a). It is achieved either by treating inside surface or outside surface.
b). The false ceiling with an air gap may be provided. The ceiling is made of thermal insulating materials.
c). The light insulating materials may be provided by suitable adhesives to the inside surfaces of the exposed roof.

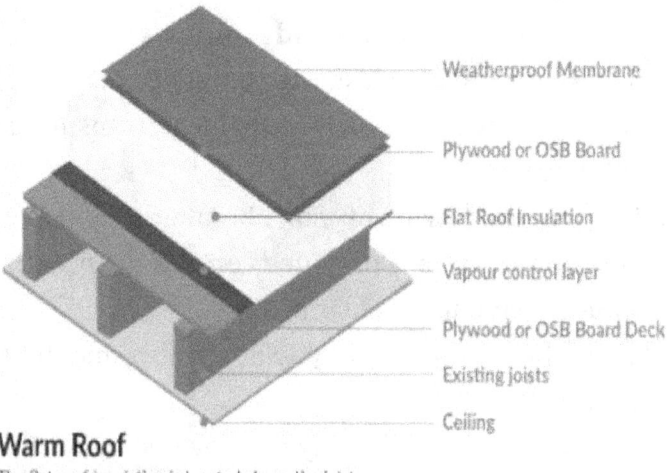

Fig. 3.18 Thermal insulation of exposed roofs

Thermal Physics

3.13 Applications

3.13.1 Heat Exchangers

A heat exchanger is a device designed to efficiently transfer or exchange heat from a hot fluid to a cold fluid with different temperatures.

Types of heat exchangers

Several types of heat exchangers have been developed which are classified on the basis of

1. Nature of heat exchange process
2. Relative direction of fluid motion
3. Design and constructional features

1. Nature of heat exchange process

It is further classified into two types

a). Direct contact heat exchangers

b). Indirect contact heat exchangers

a). Direct contact heat exchangers: It is shown in Fig. 3.19. In a direct contact or open heat exchanger, the exchange of heat takes place by direct mixing of hot and cold fluids and transfer of heat and mass takes place simultaneously.

The steam in the direct contact heat exchanger mixes with cold water and gives its latent heat to water and gets condensed. Hot water and non condensable gases leave the container.

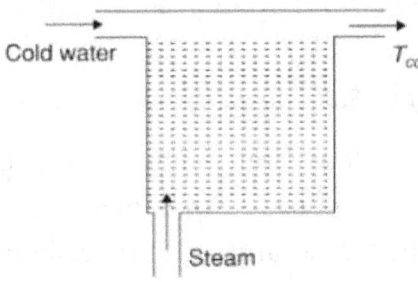

Fig. 3.19 Direct heat exchanger

Examples: Cooling towers, Jet condensers, Direct contact feat heaters

b). Indirect contact heat exchangers: In this exchanger, the heat transfer between two fluids could be carried out by transmission through wall. These exchangers are further classified into two types: (i) Regenerators, (ii) Recuperators

(i). Regenerators: The hot and cold fluids pass alternately through a space containing solid particles which provide alternately a sink and a source of heat flow.

Examples: IC engines, Air heaters and glass melting furnaces.

(ii). Recuprators: The recuperator heat exchanger in which the flowing fluids exchanging heat are on either side of dividing wall.

Examples: Automobile radiators, Milk chiller

2. Relative direction of fluid motion

According to the relative directions o fluid motions, the heat exchangers are classified into three types: (i) parallel flow or unidirectional flow, (ii) counter flow and (iii) cross flow.

(i) Parallel flow or unidirectional flow: This heat exchanger is shown in Fig. 3.20. It consists of two streams (hot and cold) which travel in same direction. The two streams enter at one end and leave the other end. The temperature difference between the hot and cold fluids decreases from inlet to outlet. Examples: oil coolers, oil heaters and water heaters.

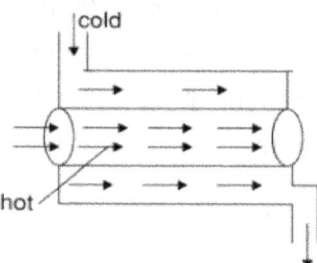

Fig. 3.20 Parallel flow

(ii) Counter flow: The counter flow arrangement of heat exchanger is shown in Fig. 3.21. Two fluids flow in opposite directions. The hot and cold fluids enter at the opposite ends. The temperature difference between the two fluids remains more or less nearly constant.

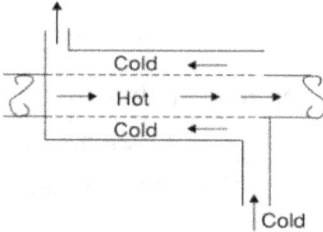

Fig. 3.21 Counter flow

Thermal Physics

(iii) Cross flow heat exchanger: Hot and cold fluids cross one another in space at right angles. It is illustrated in Fig. 3.22.

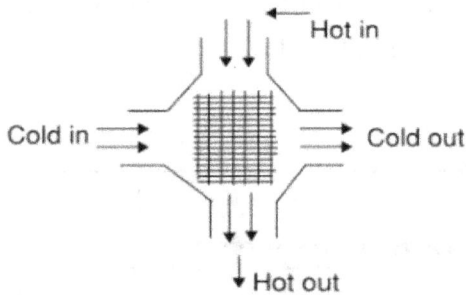

Fig. 3.22 Cross flow

3. Design and construction features

It is further classified into four types: (i) concentric tubes, (ii) shell and tube, (iii) multiple shell and tube passes and (iv) compact heat exchanger.

(i) Concentric tubes: Two concentric tubes are used to carry the fluids. The direction of flow may be parallel or counter. The effectiveness of the heat exchanger is increased by using swirl flow. The arrangement of concentric tubes is shown in Fig. 3.23.

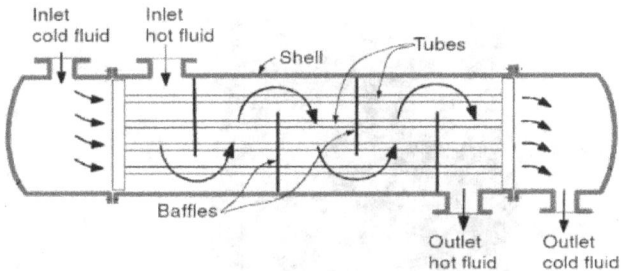

Fig. 3.23 Concentric tubes

(ii) Shell and tube: It consists of two fluids. One of the fluids flows through a bundle of tubes enclosed by a shell whereas the other fluid is forced through the shell and it flows over the outside surface of the tubes. (shown in Fig. 3.24).

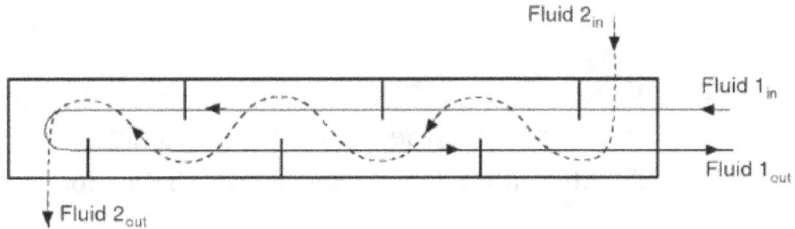

Fig. 3.24 one shell and two tube passes

(iii) Multiple shell and tube passes: They are used for enhancing the overall heat transfer. It is possible where the fluid flowing through the shell is re-routed. The shell side fluid is forced to flow back and forth across the tubes by baffle. The arrangement is shown in Fig. 3.25.

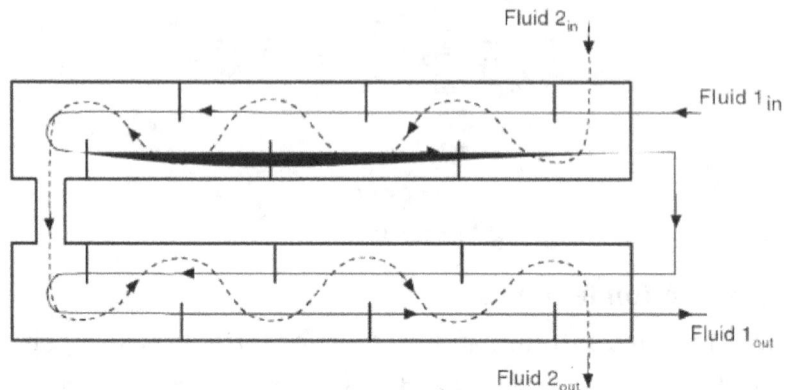

Fig. 3.25 Two shell, four tube passes

(iv) Compact heat exchangers: They have a very large transfer surface area per unit volume of the exchanger. They are generally employed when convective heat transfer coefficient associated with one of the fluids is much smaller than that associated with the other fluid.

Fig. 3.26 Compact heat exchanger

Examples: Radiator for a private car, plate fin and flattened fin tube exchangers.

3.13.2 Refrigerators

They are machines designed for the production of artificial coldness required to maintain a chamber for the storage of perishable food at low temperature.

Thermal Physics

Principle

The cooling is produced by evaporation of a liquid under reduced pressure.

Construction

It consists of the following components as shown in Fig. 3.27.

Expansion valve: It controls the flow of the liquid refrigerant into the evaporator.

Compressor: It consists of a motor that 'sucks in' the refrigerant from the evaporator and compresses (with high pressure) it in a cylinder to make a hot gas.

Evaporator: It consists of finned tubes that absorb the heat blown through a coil by a fan.

Condenser: It consists of a coiled set of tubes with external fins and is located at the rear of the refrigerator. It helps in the liquefaction of gaseous refrigerant by absorbing its heat and subsequently expelling it to the surroundings.

Refrigerant: The coolant in the refrigerator is called refrigerant. The coolant is liquid which may be ammonia, sulphur dioxide and freon (CF_2C_{12}).

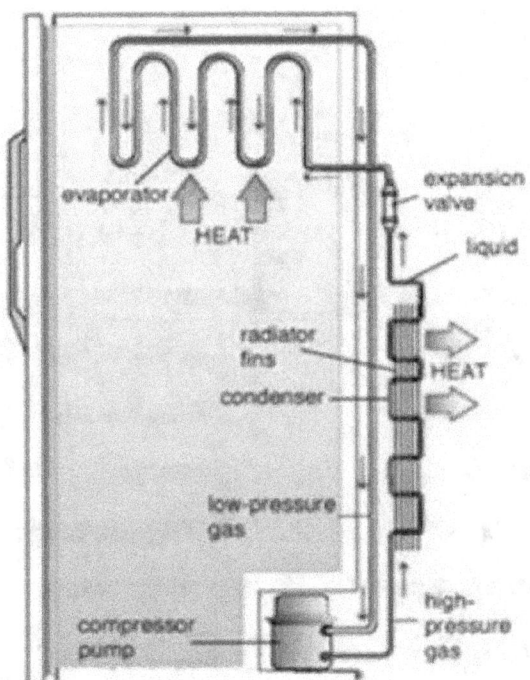

Fig. 3.27 Components of refrigerator

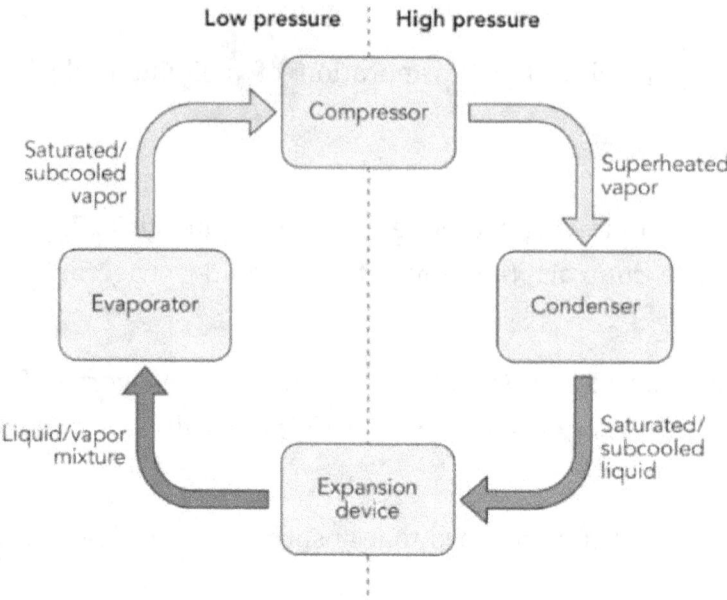

Fig. 3.28 Working flowchart of refrigerator

Working

The refrigerator schematic diagram is illustrated in Fig. 3.29. The refrigerant (liquid state) passes through the expansion valve and turns into a cool gas due to the sudden drop in pressure. When the cooled refrigerant gas flows through the chiller cabinet, it absorbs the heat from the food items inside the fridge and vaporizes.

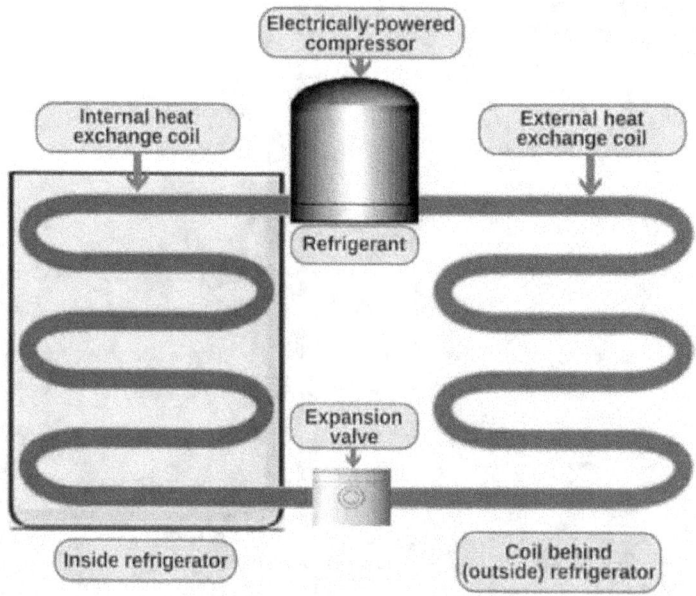

Fig. 3.29 Working of refrigerator

The refrigerant gas flows into the compressor which sucks and compresses the molecules together to make refrigerant gas as high pressure gas. The high pressure gas transports into the condenser coils.

After the condenser, the liquid refrigerant travels back to the expansion valve and then absorbs heat from the contents of the fridge and the whole cycle repeats.

3.13.3 Oven

It is thermally insulated chamber used for heating, baking or drying the substance and most commonly used for cooking.

Principle

It works on the principle of fine gravity air convection in a highly heated electrical chamber.

Description

It consists of the following parts as shown in Fig. 3.30.

a). An insulated chamber
b). A fan to circulate the air
c). Shelves
d). Thermocouples
e). Temperature sensor
f). Door locking controls

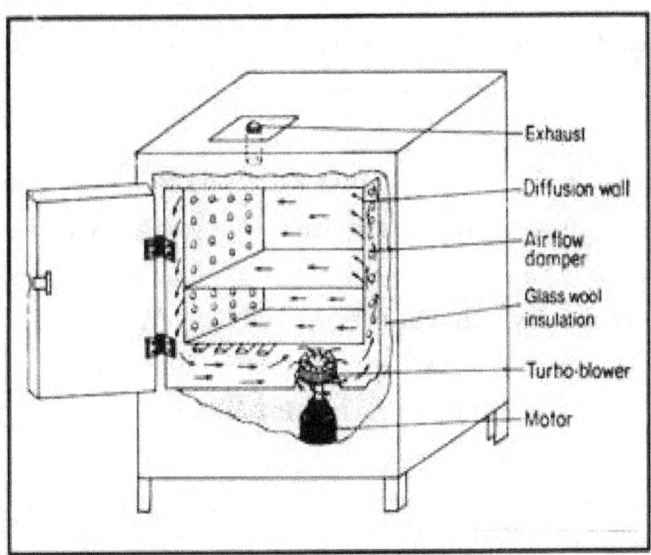

Fig. 3.30 Schematic diagram of oven

Thermal Physics

The apparatus consists of copper base and covered with asbestos sheets. The roof is provided with a hole through which a thermometer is fitted inside for recording of temperature. The oven has two or three shelves and is heated by electric heater. A regulator is used to control the temperature inside the oven.

Working

The oven is switched on, after loading of glass wares or samples. The temperature is slowly increased up to the desired point (approximately 160°C) where it will remain steady. At appropriate temperature, the oven is kept for an hour. Then the temperature is gradually brought down and thereafter sterilization is complete.

Advantages
 a). It kills the bacterial endoxin.
 b). Dry heat sterilization by hot air oven does not leave any chemical resistance.
 c). Eliminates 'wet pade' problems in humid climates.

Disadvantages
 a). Plastic and rubber items cannot be dried.
 b). Dry heat penetrates materials slowly and unevenly.
 c). It requires a continuous source of electricity.

Applications

It is widely used to sterilize glass wares in pharmaceutical industries such as petri dishes, pipettes, bottles, test tubes, etc.

3.13.4 Solar water heater

The solar water heater consists of the following parts as shown in Fig. 3.31.
 a). The solar collector, in which water is heated by solar radiation.
 b). An insulated storage tank, in which the heated water from the collector is stored. The storage tank must be put higher than the top of the collector.
 c). An insulated pipe connecting the lower part of the collector and the upper part of the storage tank.

d). An insulated pipe connecting the lower part of the storage tank and the bottom of the collector.
e). A cold water inlet connecting an existing water supply system to the storage tank. Usually the cold water inlet runs via a buffer tank with a floating gauge.
f). An insulated hot water outlet running from the storage tank to the tap.
g). A vent (air escape pipe) to prevent overpressure, caused by air or steam.

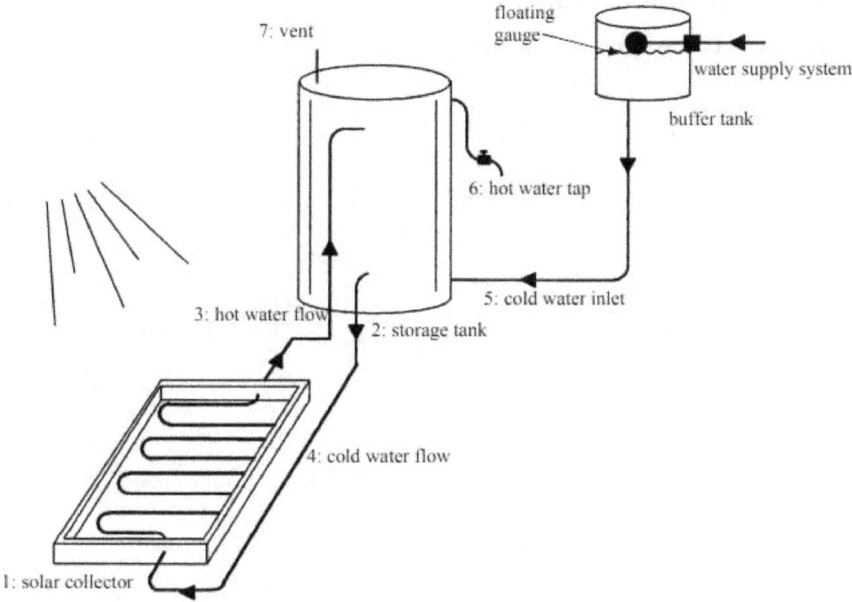

Fig. 3.31 Schematic diagram of solar water heater

Working:

When solar radiation heats the collector, the water inside will be heated as well. The heated water starts rising through the connection on top of the collector to the insulated storage tank. Heated water entering the storage tank displaces cooler water that is in turn forced via the connection to the bottom of the collector. The cold water entering the collector will be heated again by solar radiation. Because the water temperature inside the collector becomes much higher than inside the storage tank, the circulation continues as long as the sun heats the collector.

Consequently, the water inside the storage tank will get hotter and hotter. Depending on the amount of solar radiation and insulation, the system can produce water temperatures between 40 and 70°C..

Advantages
- a). It is maintenance free
- b). It does not require to add any fuel
- c). It does not release offensive smells
- d). It does not requiring costs

Disadvantages
- a). The initial cost of purchasing and installing the solar water heater is always high.
- b). It cannot be used in night time.
- c). Solar energy is to be focused at one point.

Thermal Physics

PART-A TWO MARKS

1. **Mention the types of thermal expansion of solids.**
   a). Linear expansion
   b). Superficial expansion
   c). Cubical expansion

2. **Define coefficient of linear expansion of solids.**
   The coefficient of linear expansion of a solid is increase in length produced per unit length of the solid when its temperature is raised by 1 K. It is denoted by α
   $$\alpha = \frac{L_2 - L_1}{L_1(T_2 - T_1)}$$

3. **Define coefficient of superficial expansion.**
   The coefficient of superficial expansion of a solid is increase in area produced per unit area of the solid when its temperature is raised by 1 K. It is denoted by β.
   $$\beta = \frac{A_2 - A_1}{A_1(T_2 - T_1)}$$

4. **Define coefficient of cubical expansion.**
   The coefficient of cubical expansion of a solid is increase in volume produced per unit volume of the solid when its temperature is raised by 1 K. It is denoted by υ.
   $$\upsilon = \frac{V_2 - V_1}{V_1(T_2 - T_1)}$$

5. **What is expansion joint? Mention its types.**
   It is an assembly designed to absorb the heat induced expansion or contraction of a pipeline, duct or vessel. Expansion joint is classified into two types.
   a). Metallic expansion joints
   b). Wall expansion joints

6. **What is bimetallic strip? Write its applications.**
   A bimetallic strip is known as strip made of two metals of different expansion coefficients joined together.

Bimetallic strips are used as
- a). bimetallic thermo-switches
- b). bimetallic thermostats
- c). bimetallic thermometer
- d). bimetallic sensors

7. **What are the modes of heat transferred from one place to another?**

   There are three modes of transmission of heat. They are
   - a). Conduction
   - b). Convection
   - c). Radiation

8. **Define thermal conduction or heat conduction.**

   It is known as that the heat is transferred from hot region to cold region through the substance without the motion of particles.

9. **Define the coefficient of thermal conductivity.**

   It is defined as the amount of heat conducted per second through unit area of the material when unit temperature gradient is maintained. Unit is $Wm^{-1}K^{-1}$. It is represented by K.

   $$K = \frac{Qx}{A(\theta_2 - \theta_1)t}$$

10. **List out the methods to determine thermal conductivity.**
    - a). Forbe's method for determining the absolute conductivity of metals
    - b). Lee's disc method for bad conductors
    - c). Searle's method for good conductors

11. **Explain why the specimen used to determine thermal conductivity of a bad conductor should have a larger area and smaller thickness.**

    For a bad conductor should have a larger area and smaller thickness, the amount of heat conducted increases (large). It will help to measure the heat conduction accurately.

12. **Define thermal resistance.**

    Thermal resistance is a heat property and a measurement of a temperature difference by which an object or material resists a heat flow. Thermal resistance is the reciprocal of thermal conductance.

Thermal Physics

13. How are heat conduction and electrical conduction analogous to each other?

Sl. No	Heat conduction	Electrical conduction
1.	It takes place from hot surface to cold surface	It takes place from higher potential to lower potential
2.	It is due to mainly due to free electrons	It is due to free electrons
3.	It is measured as thermal conductivity	It is measured as electrical conductivity.

14. What is meant by thermal insulation?

Thermal insulation is used to reduce the flow of heat between outside and inside of a building. Due to thermal insulation, the energy consumption is also reduced to maintain artificial cooling in summer or artificial heating in winter in a room.

15. The roof building is often painted white during summer. Why?

Due to roof building painted by white, the energy consumption is reduced to maintain artificial cooling in summer

16. What are the factors to be considered in order to maintain a comfortable inside the building?

In order to maintain a comfortable inside the building, insulating materials should

    a). have low thermal conductivity

    b). absorb less moisture

    c). have adequate fire resistance

    d). have good stability

    e). have high specific heat

    f). be available at lower cost

17. Mention the properties of the thermal insulating materials.

    a). have low thermal conductivity

    b). absorb less moisture

    c). have adequate fire resistance

    d). have good stability

    e). have high specific heat

    f). be available at lower cost

Thermal Physics

18. What is heat exchanger? Mention its types

A heat exchanger is a device designed to efficiently transfer or exchange heat from a hot fluid to a cold fluid with different temperatures. Several types of heat exchangers have been developed which are classified on the basis of

    a). Nature of heat exchange process

    b). Relative direction of fluid motion

    c). Design and constructional features

19. What is meant by solar power?

Solar power is the conversion of energy from sunlight into electricity.

20. Explain the principle of refrigeration.

They are machines designed for the production of artificial coldness required to maintain a chamber for the storage of perishable food at low temperature.

21. Define oven.

It is thermally insulated chamber used for heating, baking or drying the substance and most commonly used for cooking. It works on the principle of fine gravity air convection in a highly heated electrical chamber.

22. A rod 0.25 m long and 0.892×10^{-4} m² area of cross section is heated at one end through 393 K while the other end is kept at 323 K. The quantity of heat which will flow in 15 minutes along the rod is 8.811×10^3 joule. Calculate thermal conductivity of rod.

Given:

    A = 0.892×10^{-4} m²; x = 0.25 m; $\theta_1 - \theta_2$ = 393 K − 323 K = 70 K

    Q = 8.811×10^3 J; t = 15 minutes = 15×60 = 900 s

Thermal conductivity, $\quad K = \dfrac{Qx}{A(\theta_2 - \theta_1)t}$

$$K = \dfrac{8.811 \times 10^3 \times 0.25}{0.892 \times 10^{-4} \times 70 \times 900}$$

Thermal conductivity, K = 392 Wm⁻¹K⁻¹

23. The total area of the glass window is 0.5 m². Calculate how much heat is conducted per hour through the glass window if thickness of the glass is 7×10⁻³ m. The temperature of the inside surface is 25°C and of the outside surface is 40°C. Thermal conductivity of glass is 1.0 Wm⁻¹K⁻¹.

Given: A = 0.5 m²; θ_1 = 40°C = 273+40 = 313 K

θ_2 = 25°C = 273+25 = 298 K.

K = 1 Wm⁻¹K⁻¹; t = 1 hour = 60×60 = 3600 s

Amount of heat conducted, $\quad Q = \dfrac{KA(\theta_1 - \theta_2)t}{x}$

$$Q = \dfrac{1 \times 0.5 \times (313 - 298) \times 3600}{7 \times 10^{-3}} = \dfrac{27000}{7 \times 10^{-3}}$$

Q = 3857 × 10³ J = 3.857×10⁶ J

PART-B

1. Define expansion of joints. What are the types of expansion joints and write in detail about it.
2. Write the principle and working of bimetallic strip. Mention the application and advantages of bimetallic strips.
3. Describe Forbe's method to determine thermal conductivity of metals with relevant theory and experiment.
4. Describe Lee's disc method for determining thermal conductivity of bad conductors.
5. Derive the expression for effective thermal conductivity through compound media in series and parallel.
6. Write an essay about thermal insulation inside a room.
7. What is heat exchanger? Explain in detail about heat exchangers.
8. Explain the construction and working of refrigerator with neat diagram.
9. Explain the construction and working of oven with neat diagram.
10. Explain the construction and working of solar water heater with neat diagram.

UNIT-IV
QUANTUM PHYSICS

4.1 Black body radiation

A perfect black body is one which absorbs the entire radiations (all the wavelengths) incident on it and when a body is placed at constant high temperature, it emits radiation of all the wavelengths. The heat radiation emitted from the black body is known as black body radiation.

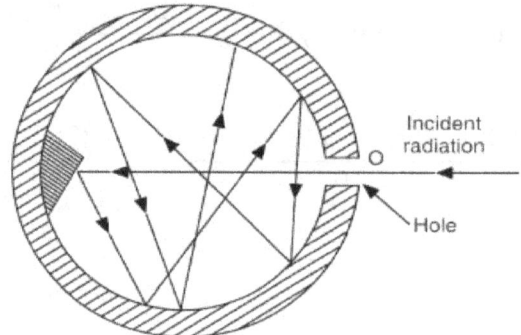

Fig. 4.1 Black body radiation

Laws of black body radiation

i). Wein's displacement law: It states that the product of wavelength (λ_m) and absolute temperature of the black body (T) is a constant.

$$i.e. \lambda_m T = \text{constant}$$

This law holds good only for shorter wavelength and not for longer wavelength.

ii) Rayleigh-Jeans law: It states that the energy distribution of a black body is directly proportional to the absolute temperature (T) and inversely proportional to the fourth power of the wavelength (λ).

$$i.e. \ E_\lambda = \frac{8\pi k_B T}{\lambda^4}$$

This law holds good only for longer wavelength and not for shorter wavelength.

Quantum Physics

iii) Stefan-Boltzmann law: The heat energy (E) radiated per unit area of a black body is directly proportional to the fourth power of its absolute temperature.

i.e. $E \alpha T^4$

$$E = \sigma T^4$$

4.2 Planck's quantum theory (derivation)

Assumptions:

a). A black body radiator contains a large number of oscillating particles. They are called simple harmonic oscillators which are capable of vibrating with all possible frequencies.

b). The frequency of radiation emitted by an oscillator is same as that of the frequency of vibration.

c). The oscillators radiate energy in discrete as shown in Fig. 4.2 and not in as continuous.

d). The oscillators exchange energy in the form of either absorption or emission in terms of $h\nu$. i.e $E = nh\nu$ where $n = 0, 1, 2, 3 \ldots \ldots \ldots n$.

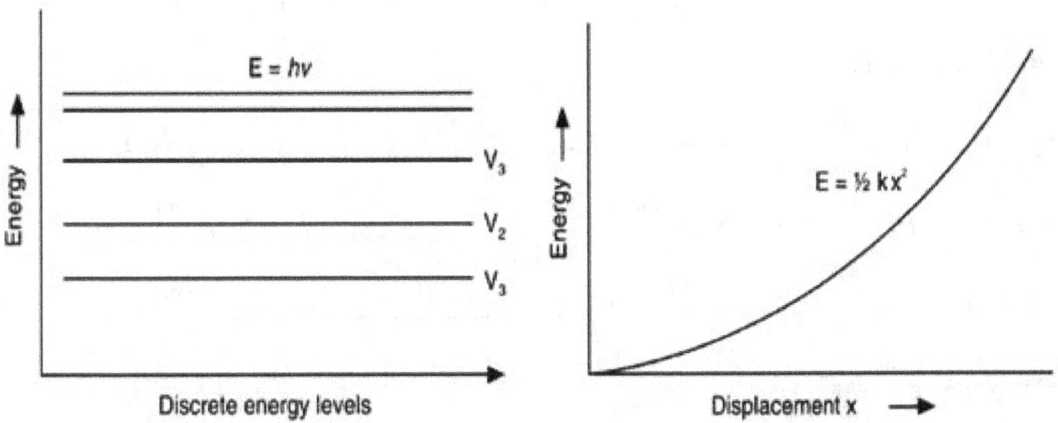

Fig. 4.2 Energy vs discrete energy levels

Derivation

Consider 'N' number of oscillations with their total energy as E_T. Then the average energy of an oscillator is given by,

$$\bar{E} = \frac{E_T}{N} \quad (1)$$

Total number of oscillations

$$N = N_0 + N_1 + N_2 + N_3 + \ldots\ldots\ldots\ldots +N_n \quad (2)$$

Total energy of oscillations

$$E_T = E_0 N_0 + E_1 N_1 + E_2 N_2 + \ldots\ldots\ldots\ldots +E_n N_n \quad (3)$$

According to Maxwell's distribution law, the number of oscillators having an energy (E_n) is given by,

$$N_n = N_0 e^{-E_n/k_B T} \quad (4)$$

From the above equation

$$N_0 = N_0 \quad \text{for } n = 0$$

$$N_1 = N_0 e^{-E_1/k_B T} \quad \text{for } n = 1$$

$$N_2 = N_0 e^{-E_2/k_B T} \quad \text{for } n = 2$$

$$N_n = N_0 e^{-E_n/k_B T} \quad \text{for } n = n$$

Eq (2) becomes as,

$$N = N_0 + N_0 e^{-E_1/k_B T} + N_0 e^{-E_2/k_B T} + \ldots\ldots\ldots +N_0 e^{-E_n/k_B T}$$

$$N = N_0 + N_0 e^{-h\upsilon/k_B T} + N_0 e^{-2h\upsilon/k_B T} + \ldots\ldots\ldots +N_0 e^{-nh\upsilon/k_B T}$$

$$N = N_0 (1 + e^{-h\upsilon/k_B T} + e^{-2h\upsilon/k_B T} + \ldots\ldots\ldots +e^{-nh\upsilon/k_B T})$$

Take, $x = e^{-h\upsilon/k_B T}$

$$N = N_0 (1 + x + x^2 + \ldots\ldots\ldots\ldots +x^n)$$

$$N = \frac{N_0}{1 - x} \quad (5)$$

Similarly, equation (3) becomes as

$$E_T = E_0 N_0 + E_1 N_0 e^{-E_1/k_B T} + E_2 N_0 e^{-E_2/k_B T} + \ldots\ldots +E_n N_0 e^{-E_n/k_B T}$$

$$E_T = E_0 N_0 + h\upsilon N_0 e^{-h\upsilon/k_B T} + 2h\upsilon N_0 e^{-2h\upsilon/k_B T} + \ldots\ldots +nh\upsilon N_0 e^{-nh\upsilon/k_B T}$$

$$E_T = E_0 N_0 + h\upsilon N_0 x + 2h\upsilon N_0 x^2 + \ldots\ldots\ldots + nh\upsilon N_0 x^n$$

$$E_T = h\upsilon N_0 x(1 + 2x + \ldots\ldots\ldots + nx^{n-1})$$

$$E_T = \frac{h\upsilon N_0 x}{(1-x)^2} \tag{6}$$

Substituting equations (5) and (6) in equation (1)

$$\overline{E} = \frac{\frac{h\upsilon N_0 x}{(1-x)^2}}{\frac{N_0}{1-x}}$$

$$\overline{E} = \frac{h\upsilon x}{1-x}$$

$$\overline{E} = \frac{h\upsilon}{\frac{1}{x} - 1}$$

$$\overline{E} = \frac{h\upsilon}{\frac{1}{e^{-h\upsilon/k_B T}} - 1}$$

$$\overline{E} = \frac{h\upsilon}{e^{h\upsilon/k_B T} - 1} \tag{7}$$

Energy density = No of oscillators per unit volume × Average energy of an oscillator

$$E_\upsilon = \frac{8\pi\upsilon^2}{c^3} d\upsilon \times \frac{h\upsilon}{e^{h\upsilon/k_B T} - 1}$$

$$E_\upsilon = \frac{8\pi h\upsilon^3}{c^3} d\upsilon \times \frac{1}{e^{h\upsilon/k_B T} - 1} \tag{8}$$

Equation (8) represents the Planck's radiation law in terms of frequency.

We know $\upsilon = \frac{c}{\lambda}$ & $d\upsilon = -\frac{c}{\lambda^2} d\upsilon$ and numerically $|d\upsilon| = \left|\frac{c}{\lambda^2}\right| d\lambda$

Equation (8) can be written as,

$$E_\lambda = \frac{8\pi hc^3}{c^3 \lambda^3} \frac{c}{\lambda^2} d\lambda \times \frac{1}{e^{hc/\lambda k_B T} - 1}$$

$$E_\lambda = \frac{8\pi hc}{\lambda^5} d\lambda \times \frac{1}{e^{hc/\lambda k_B T} - 1} \qquad (9)$$

Eq (9) represents the Planck's radiation law in terms of wavelength.

Case (i) deduction of Wein's displacement law

When λ is small, υ is large. So equation (9) can be written as

$$E_\lambda = \frac{8\pi hc}{\lambda^5} d\lambda \times \frac{1}{e^{hc/\lambda k_B T}} \qquad (10)$$

Case (ii) deduction of Rayleigh-Jeans law

When λ is large, υ is small. So equation (9) can be written as

$$E_\lambda = \frac{8\pi k_B T}{\lambda^4} \qquad (11)$$

4.3 Compton effect (Theory and experiment verification)

When a beam of high frequency radiation (such as x-rays) is scattered by a substance of low atomic number, the scattered radiation consists of two components. One has the same wavelength (λ) as the incident ray and the other component has a slightly longer wavelength (λ'). The change in the wavelength of scattered x-rays is known as Compton shift. The phenomenon is called Compton effect.

Consider an x-ray photon striking an electron at rest as shown in Fig. 4.3(a). This x-ray photon is scattered through an angle 'θ' to x-axis from its initial direction of motion.

Let the frequency of scattered photon be υ' and its energy $h\upsilon'$. During the collision, x-ray photon gives a fraction of its energy to the free electron. This free electron of mass 'm' gains energy and it moves with a velocity 'v' at an angle ϕ to x-axis.

Quantum Physics

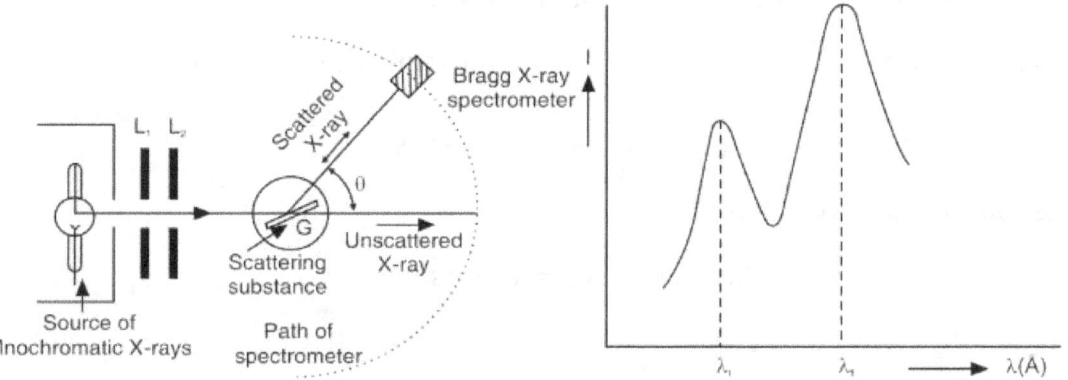

(a) Set up of Compton effect   (b) Unmodified and modified components

Fig. 4.3 Compton effect

Energy along x-axis

Total energy before collision along x − axis = $h\upsilon + m_0 c^2$ (1)

Total energy after collision along y − axis = $h\upsilon + m_0 c^2$ (2)

According to the law of conservation of energy,

     Energy before collision = Energy after collision

$$h\upsilon + m_0 c^2 = h\upsilon + m_0 c^2 \quad (3)$$

Momentum along x-axis

Total momentum before collision along x − axis = $\dfrac{h\upsilon}{c}$ (4)

Total momentum after collision along x − axis = $\dfrac{h\upsilon'}{c}\cos\theta + mv\cos\phi$ (5)

According to the law of conservation of momentum,

Total momentum before collision = Total momentum after collision

$$\dfrac{h\upsilon}{c} = \dfrac{h\upsilon'}{c}\cos\theta + mv\cos\phi \quad (6)$$

Momentum along y-axis

Total momentum before collision along y − axis = 0   (7)

Total momentum after collision along x − axis = $-\dfrac{h\upsilon'}{c}\sin\theta + mv\sin\phi$ (8)

According to the law of conservation of momentum,

Total momentum before collision = Total momentum after collision

$$0 = -\frac{h\upsilon'}{c}\sin\theta + mv\sin\phi \tag{9}$$

Equation (6) can be written as

$$h(\upsilon - \upsilon'\cos\theta) = mv\cos\phi \tag{10}$$

Equation (9) can be written as

$$h\upsilon'\sin\theta = mv c\sin\phi \tag{11}$$

Squaring equations (4) and (5) and adding them, we get

$$h^2(\upsilon^2 + \upsilon'^2 - 2\upsilon\upsilon'\cos\theta) = m^2v^2c^2 \tag{12}$$

From eq(3), $\quad h(\upsilon - \upsilon') + m_0 c^2 = mc^2 \tag{13}$

Squaring the equation (13),

$$h^2(\upsilon^2 - 2\upsilon\upsilon' + \upsilon'^2) + 2h(\upsilon - \upsilon')m_0 c^2 + m_0^2 c^4 = m^2 c^4 \tag{14}$$

Subtracting equation (12) from equation (14), we get

$$m^2 c^2(c^2 - v^2) = -2h^2\upsilon\upsilon'(1 - \cos\theta) + 2h(\upsilon - \upsilon')m_0 c^2 + m_0^2 c^4 \tag{15}$$

We know that

$$m = \frac{m_0}{\sqrt{1 - \frac{v^2}{c^2}}}$$

From the above equation,

$$m^2 c^2(c^2 - v^2) = m_0^2 c^4 \tag{16}$$

Substituting equation (16) in equation (15), we get

$$m^2 c^2(c^2 - v^2) = -2h^2\upsilon\upsilon'(1 - \cos\theta) + 2h(\upsilon - \upsilon')m_0 c^2 + m^2 c^2(c^2 - v^2)$$

$$0 = -2h^2\upsilon\upsilon'(1 - \cos\theta) + 2h(\upsilon - \upsilon')m_0 c^2$$

$$2h^2\upsilon\upsilon'(1 - \cos\theta) = 2h(\upsilon - \upsilon')m_0 c^2$$

$$h\upsilon\upsilon'(1 - \cos\theta) = (\upsilon - \upsilon')m_0 c^2$$

$$\frac{h}{m_0 c^2}(1-\cos\theta) = \frac{(\upsilon - \upsilon')}{\upsilon\upsilon'}$$

$$\frac{h}{m_0 c^2}(1-\cos\theta) = \frac{1}{\upsilon'} - \frac{1}{\upsilon}$$

$$\frac{h}{m_0 c}(1-\cos\theta) = \frac{c}{\upsilon'} - \frac{c}{\upsilon}$$

$$\frac{h}{m_0 c}(1-\cos\theta) = \lambda' - \lambda$$

$$d\lambda = \frac{h}{m_0 c}(1-\cos\theta) \qquad (17)$$

Equation (17) represents the change in wavelength or Compton wavelength.

Special cases:

Case (i). When $\theta=0°$, $\cos 0° = 1$, Hence $d\lambda = 0$

Case (ii), When $\theta=90°$, $\cos 90° = 0$, Hence $d\lambda = 0.0243$.

The shift or change in wavelength is called Compton wavelength or Compton shift.

Case (iii) When $\theta = 180°$, $\cos 180° = -1$ Hence $d\lambda = 0.0485$.

The change in wavelength is maximum at the scattering angle of $\theta = 180°$

4.3.1 Experimental verification:

A beam of monochromatic X-rays of wavelength (λ) is made to fall on a scattering material. The wavelength and intensity of the scattered X-rays received by the spectrometer is measured for various values of scattering angles.

Graphs are plotted between the intensity of the scattered X-rays and wavelengths for various scattering angles. The change in wavelength ($d\lambda = 0.0243$) at $\theta = 90°$ is found to be in good agreement with the theoretical value. Thus Compton effect is experimentally verified

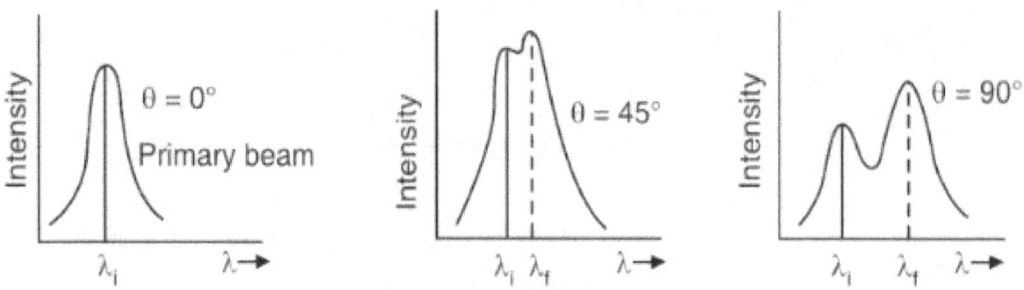

Fig. 4.4 Experimental verification of Compton effect

4.4 Wave particle duality

When the particles are in motion, they exhibit wave nature. It is known as particle duality

The waves associated with these particles are known as matter waves or de-Broglie waves.

Properties of matter waves

a). If m is large or υ is large, the wavelength associated with the particle is small.

b). The matter waves are generated when the particles are in motion.

c). The wavelength associated with a particle is independent of any charge associated with it.

4.5 Electron diffraction: G. P. Thomson experiment
Description

The experimental set up of G.P. Thomson experiment is shown in Fig. 4.5. The electrons are produced from a heat filament (F). The electron beam passes through a fine hole in a metal block (B) and falls on the surface of a thin coil (G). A photographic plate (P) is kept at a short distance from the gold foil. The whole apparatus is exhausted to a high vaccum.

Working

When a beam of electrons is passed through a thin gold foil, the electrons are diffracted. The diffracted pattern of electrons is made to fall on photographic plate (P). When the plate is developed, a symmetrical pattern consisting of concentric rings around a central spot is obtained

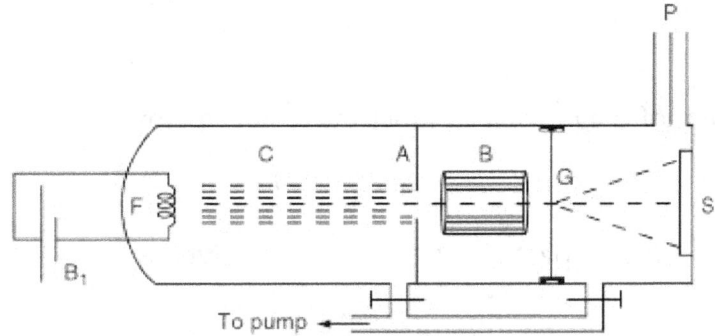

Fig. 4.5 G. P. Thomson experiment

Result

The wavelengths obtained by diffracted pattern agreed well with the wavelength obtained by the de-Braglie relation.

4.6 Concepts of wave function

The variable quantity that characterizes the matter waves is called wave function. It is represented by 'Ψ'.

Physical significance of wave function (Ψ)

a). The wave function connects the particle and wave nature statistically.

b). It relates the probability of finding the particle at the point at time.

c). The wave function (Ψ) is a complex quantity and it does not have any physical meaning by itself.

d). The probability of finding the particle in volume (dτ) is given by $\iiint \Psi^* \Psi d\tau = \iiint |\Psi|^2 d\tau$ where dτ = dxdydz.

e). The probability density is given by the equation $p(\bar{r}, t) = |\Psi(\bar{r}, t)|^2$

4.7 Schroedinger wave equations

4.7.1 Schroedinger time dependent wave equation

A particle can behave as a wave only under motion. So it should be accelerated by a potential energy.

$$E = V + \frac{1}{2}mv^2$$

$$E = V + \frac{1}{2}\frac{m^2 v^2}{m}$$

$$E = V + \frac{p^2}{2m}$$

$$E\Psi = V\Psi + \frac{p^2}{2m}\Psi \qquad (1)$$

The displacement of the particle at any time 't' is given by

$$y = Ae^{-i\omega(t-x/v)} \qquad (2)$$

The position of a moving particle at any time 't' is given by

$$\Psi_{(x,y,z,t)} = Ae^{-i\omega(t-x/v)}$$

$$\Psi_{(x,y,z,t)} = Ae^{-i2\pi(\upsilon t - \upsilon x/v)} \qquad (3)$$

We know that

$$E = h\upsilon \text{ or } \upsilon = \frac{E}{h} \qquad (4)$$

$$\upsilon = \frac{v}{\lambda} \text{ or } \frac{\upsilon}{v} = \frac{1}{\lambda} \qquad (5)$$

Substituting equations (4) and (5) in equation,

$$\Psi_{(x,y,z,t)} = Ae^{-i2\pi(Et/h - x/\lambda)} \qquad (6)$$

We know that

$$\lambda = \frac{h}{mv} = \frac{h}{p} \qquad (7)$$

Substituting equation (7) in equation (6)

$$\Psi_{(x,y,z,t)} = Ae^{-i2\pi(Et/h - px/h)}$$

$$\Psi_{(x,y,z,t)} = Ae^{-\frac{i}{\hbar}(Et - px)} \qquad (8)$$

where $\hbar = \frac{h}{2\pi}$

Differentiating equation (8) with respect to x,

$$\frac{d\Psi}{dx} = Ae^{-\frac{i}{\hbar}(Et-px)} \cdot \left(\frac{ip}{\hbar}\right)$$

Differentiating the above equation with respect to x again, we get

$$\frac{d\Psi}{dx} = Ae^{-\frac{i}{\hbar}(Et-px)} \cdot \left(\frac{ip}{\hbar}\right)^2$$

$$\frac{d\Psi}{dx} = \Psi \cdot \left(\frac{ip}{\hbar}\right)^2$$

$$p^2\Psi = -\hbar^2 \frac{d^2\Psi}{dx^2} \qquad (9)$$

Differentiating equation (8) with respect to 't'

$$\frac{d\Psi}{dt} = Ae^{-\frac{i}{\hbar}(Et-px)} \cdot \left(\frac{-iE}{\hbar}\right)$$

$$\frac{d\Psi}{dx} = \frac{E\Psi}{i\hbar}$$

$$E\Psi = i\hbar \frac{d\Psi}{dx} \qquad (10)$$

Substituting equation (9) and equation (10) in equation (1), we get

$$i\hbar \frac{d\Psi}{dx} = V\Psi - \frac{\hbar^2}{2m} \frac{d^2\Psi}{dx^2}$$

$$i\hbar \frac{d\Psi}{dx} = \left[V - \frac{\hbar^2}{2m}\frac{d^2}{dx^2}\right]\Psi \qquad (11)$$

Equation (11) represents the one dimensional Schroedinger time dependent wave equation.

Schroedinger Time independent wave equation

We know that
$$\Psi_{(x,y,z,t)} = Ae^{-\frac{i}{\hbar}(Et-px)}$$

$$\Psi_{(x,y,z,t)} = Ae^{-\frac{iEt}{\hbar}} \cdot e^{\frac{ipx}{\hbar}}$$

$$\Psi_{(x,y,z,t)} = A\Psi e^{-\frac{iEt}{\hbar}} \qquad (1)$$

where $\Psi = e^{-\frac{ipx}{\hbar}}$ and represents the time independent wave equation.

Differentiating equation with respect to 'x',

$$\frac{d\Psi}{dx} = Ae^{-\frac{iEt}{\hbar}} \cdot \frac{d\Psi}{dx}$$

Differentiating the above equation with respect to 'x',

$$\frac{d^2\Psi}{dx^2} = Ae^{-\frac{iEt}{\hbar}} \cdot \frac{d^2\Psi}{dx^2} \qquad (2)$$

Differentiating equation (1) with respect to 't',

$$\frac{d\Psi}{dt} = A\Psi e^{-\frac{iEt}{\hbar}} \cdot \left(\frac{-iE}{\hbar}\right) \qquad (3)$$

Schroedinger Time independent wave equation is

$$i\hbar \frac{d\Psi}{dt} = V\Psi - \frac{\hbar^2}{2m}\frac{d^2\Psi}{dx^2} \qquad (4)$$

Substituting equations (1), (2) and (3) in equation (4), we get

$$i\hbar A\Psi e^{-\frac{iEt}{\hbar}} \cdot \left(\frac{-iE}{\hbar}\right) = V\Psi - \frac{\hbar^2}{2m} Ae^{-\frac{iEt}{\hbar}} \cdot \frac{d^2\Psi}{dx^2}$$

$$i\hbar \left(\frac{-iE}{\hbar}\right)\Psi = V\Psi - \frac{\hbar^2}{2m}\frac{d^2\Psi}{dx^2}$$

$$(-i)^2 E\Psi = V\Psi - \frac{\hbar^2}{2m}\frac{d^2\Psi}{dx^2}$$

$$E\Psi - V\Psi = -\frac{\hbar^2}{2m}\frac{d^2\Psi}{dx^2}$$

$$\frac{d^2\Psi}{dx^2} = -\frac{2m}{\hbar^2}(E\Psi - V\Psi)$$

$$\frac{d^2\Psi}{dx^2} + \frac{2m}{\hbar^2}(E - V)\Psi = 0 \qquad (5)$$

Equation (5) represents the one dimensional Schroedinger time independent wave equation.

Quantum Physics

4.8 Particle in a one dimensional rigid box

Let us consider a particle of mass 'm' moving along x-axis enclosed in a one dimensional potential box as shown in Fig. 4.6. The walls are infinite potential. The particle does not penetrate out from the box. The boundary condition is

$$v(x) = 0 \text{ when } 0 < x < l$$
$$v(x) = \alpha \text{ when } 0 \geq x \geq l$$

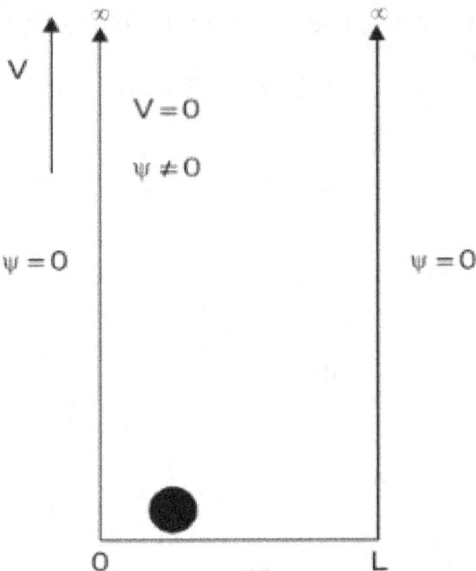

Fig. 4.6 Particle in one dimensional potential box

Schroedinger time independent wave equation is

$$\frac{d^2\Psi}{dx^2} + \frac{2m}{\hbar^2}(E - V)\Psi = 0$$

Since the potential energy inside the wall is zero, the above equation becomes

$$\frac{d^2\Psi}{dx^2} + \frac{2m}{\hbar^2}E\Psi = 0$$

$$\frac{d^2\Psi}{dx^2} + k^2\Psi = 0 \qquad (1)$$

where
$$k^2 = \frac{2mE}{\hbar^2} \qquad (2)$$

The solution of equation (1) is

$$\Psi(x) = A\sin kx + B\cos kx \qquad (3)$$

Where A and B are arbitrary constants

Condition (i): At $x = 0$ and $v = \alpha$, there is no particle at the walls of the box.

$$\therefore \Psi(x) = 0$$

Equation (3) becomes, $B = 0$

Condition (ii): At $x = l$ and $v = \alpha$, there is no particle at the walls of the box.

$$\therefore \Psi(x) = 0$$

Equation (3) becomes, $\quad A\sin kl = 0$

$$\sin kl = 0 \qquad \text{where } A \neq 0$$

We know that $\quad \sin n\pi = 0$

$$\therefore n\pi = kl$$

$$k = \frac{n\pi}{l} \qquad (4)$$

Substituting $B = 0; k = \dfrac{n\pi}{l}$ in equation (3), we get

$$\Psi(x) = A\sin\frac{n\pi x}{l} \qquad (5)$$

Equation (5) represents the wave function of a particle in one dimensional box.

Energy of the particle:

Equation (2) can be written as $\quad k^2 = \dfrac{2mE}{h^2/4\pi^2}$

$$k^2 = \frac{8\pi^2 mE}{h^2} \qquad (6)$$

from equation (4),
$$k^2 = \frac{n^2\pi^2}{l^2} \qquad (7)$$

Equating equations (6) and (7),

$$\frac{8\pi^2 mE}{h^2} = \frac{n^2\pi^2}{l^2}$$

$$E = \frac{n^2\pi^2 h^2}{8\pi^2 m l^2}$$

Enegry of the particle,
$$E_n = \frac{n^2 h^2}{8ml^2} \qquad (8)$$

E_n is said to be Eugen value corresponding to the Eigen function (Ψ_n).

Energy levels of the electrons

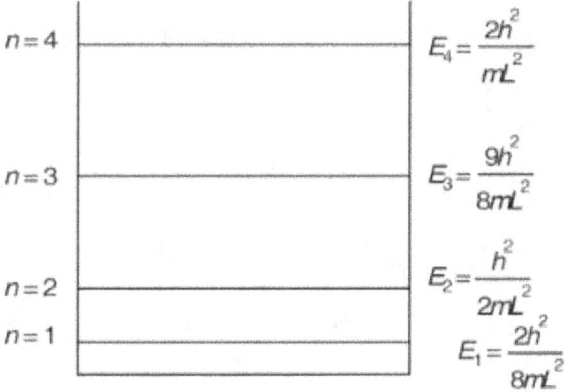

Fig. 4.7 Energy levels of the electrons

For $n = 1$, $E_1 = \dfrac{h^2}{8ml^2}$

For $n = 1$, $E_2 = \dfrac{4h^2}{8ml^2} = 4E_1$

For $n = 1$, $E_3 = \dfrac{9h^2}{8ml^2} = 9E_1$

In general, $E_n = n^2 E_1 \qquad (9)$

Quantum Physics

Normalization of the wave function:

Probability of one dimensional box of the length (l) is given by

$$\int_0^l |\Psi|^2 dx = 1 \tag{10}$$

Substituting equation (5) in equation (10),

$$\int_0^l A^2 \sin^2 \frac{n\pi x}{l} dx = 1$$

$$A^2 \int_0^l \frac{1 - \cos \frac{2n\pi x}{l}}{l} dx = 1$$

$$A^2 \left[\frac{l}{2} - \frac{1}{2} \frac{\sin 2\pi n}{2\pi n / l} \right] = 1$$

We know that $\sin 2\pi n = 0$, so $\quad \dfrac{A^2 l}{2} = l$

$$A^2 = \frac{2}{l}$$

$$A = \sqrt{\frac{2}{l}}$$

Substituting the value of A in equation (5),

$$\Psi(x) = \sqrt{\frac{2}{l}} \sin \frac{n\pi x}{l} \tag{11}$$

Equation (11) is said to be normalized wave function.

Quantum Physics

4.9 Tunneling

Tunneling can be defined as that there is a finite chance for the particle to penetrate through the potential barrier and escape from the other side of potential barrier.

Applications

a). The tunnel side is a semiconductor diode in which electrons tunnel through a potential barrier.

b). The current can be switched on and off very quickly by varying the height of the barrier.

c). The scanning tunneling electron microscope uses electron tunneling to produce images of surfaces down to the scale of individual atoms.

4.10 Scanning Tunneling Microscope (STM)

In 1981, Gerd Binning and Heinrich Rohrer developed the scanning tunneling microscope for observing surfaces of the materials atom by atom.

Principle

STM has a metal needle which scans a sample by moving back and forth. The needle does not touch the sample but stays about the width of two atoms above its surface.

Instrumentation

A schematic of STM is shown in Figure. It has the following components.

Piezoelectric tube: It is capable of moving in X, Y, Z direction.

Fine needle tip: It is for scanning the sample surface.

Tip is affixed to the piezoelectric tube to control its position and movement on an atomic scale. Piezoelectrical materials exhibit an elongation or contraction along their length when an electric field is applied.

Applying voltages to the appropriate regions on the tube causes the piezoelectric tube to deform and move the tip. The X and Y piezos control the back and forth motion of the tip while the Z-piezo allow the tip to approach the surface.

Working

The tip is connected to the scanner. The sharp metal needle is brought close to the surface to be imaged. A bias voltage is applied between the sample and the tip. When the needle is at positive potential, electrons can be tunned through the gap and set up a small tunneling current in the needle. This tunneling current is amplified and measured.

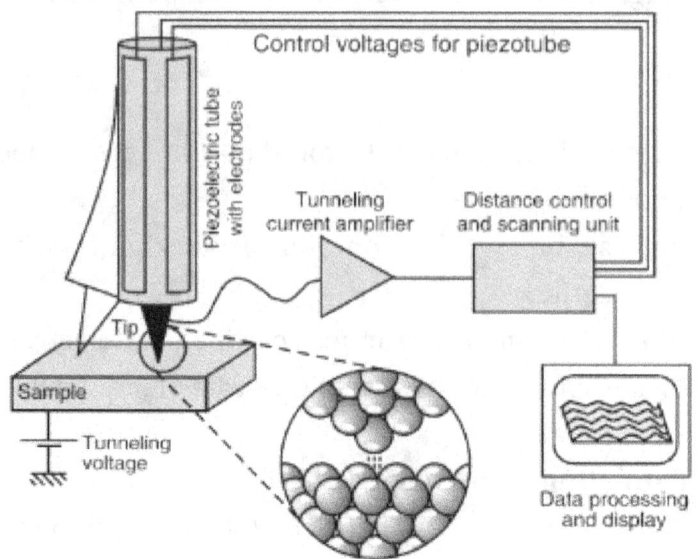

Fig. 4.8 Schematic diagram of scanning tunneling microscope

Applications

a). The STM shows the position of atoms

b). It provides non-trivial real space information in studying semiconductor

c). It is used to manipulate the atoms

d). To analyze the electronic structures of the active sites at catalyst surfaces

Quantum Physics

PART-A TWO MARKS

1. **Define blackbody and blackbody radiation.**

    A perfect blackbody is one which absorbs and also emits the radiations of all the wavelengths. Blackbody is said to be a perfect absorber, since it absorbs all the wavelength of the incident radiation. The blackbody is a perfect radiator, because it radiates the entire wavelength absorbed by it. This phenomenon is also called blackbody radiation

2. **What is meant by energy spectrum of a black body? What do you infer from it?**

    The energy spectrum of black body can be obtained by plotting a curve between the intensity of radiation along Y axis and wavelength along X axis at different temperatures. The characteristics of black body can be studied using this plot.

3. **What are the postulates of Planck's quantum theory?**
    a). The frequency of radiation emitted by an oscillator is the same as that of the frequency of vibration.
    b). The oscillators radiate energy in discrete and not in continuous manner
    c). The oscillators exchange energy in the form of either absorption or emission in terms of quantum of photon energy 'hυ'
    d). A black body radiator contains a large number of oscillating particles which are capable of vibrating with all the frequencies.

4. **Write Planck's radiation formula.**

    The energy density of heat radiation emitted from a blackbody at temperature (T) in the wavelength range from λ to λ+dλ is given by

    $$E_\lambda d\lambda = \frac{8\pi hc}{\lambda^5 (e^{\frac{h\upsilon}{kT}} - 1)}$$

    Where h-Plnack's constant, c-speed of the light, v-frequency of radiation, k-Boltzman constant.

5. **State Wien's displacement law of black body radiation.**

    The product of wavelength (λ_m) corresponding to the maximum energy of radiation and absolute temperature of the body (T) is a constant. $\lambda_m T$ = constant.

Quantum Physics

6. What is Rayleigh-Jeans law of radiation?

The energy distribution of a blackbody is directly proportional to the absolute temperature (T) and inversely proportional to the fourth power of the wavelength (λ).

$$E_\lambda = \frac{8\pi k_B T}{\lambda^4}$$

7. What is Compton effect? Write an expression for the Compton wavelength.

When a beam of monochromatic radiation of high frequency is scattered by a scattering element, the scattered beam consists of two components, one component having the same frequency and other component has a slightly lower frequency as that of the incident radiations. The change in the wavelength of scattered beam is known as Compton shift.

$$d\lambda = \frac{h}{m_0 c}(1 - \cos\theta)$$

8. What is Compton wavelength or Compton shift? Give its value.

The change in wavelength corresponding to the scattering angle of 90° is called Compton wavelength.

$$d\lambda = \frac{h}{m_0 c}(1 - \cos\theta)$$

when $\theta = 90°$, $\cos 90° = 0$, $d\lambda = 0.0242$ A°

9. What is meant by matter wave? How are matter-waves different from electromagnetic waves?

A particle associated with its wave nature is known as matter waves. They do not depend on the charge of the particles. It shows that these waves are not electromagnetic waves.

10. Write down the expression for the wavelength of matter waves.

The wavelength of matter wave or de-Broglie wave associated with any moving particle of mass 'm' with velocity 'v' is given by

$$\lambda = \frac{h}{mv} = \frac{h}{p}$$

De-Broglie wavelength in terms of energy:

$$\lambda = \frac{h}{\sqrt{2mE}}$$

De-Broglie wavelength in terms of potential difference (V):

$$\lambda = \frac{h}{\sqrt{2m_e V}}$$

11. List out the properties of matter waves

a). If the mass 'm' or frequency 'υ' of the particle is large, then the wavelength associated with the particle is small.

b). The matter waves are generated when the particles are in motion.

c). The matter waves do not dependent on the charged particles.

12. Write down the Schrodinger wave equations.

a). Schrodinger time dependent wave equation

$$i\hbar \frac{d\Psi}{dt} = \left[V - \frac{\hbar^2}{2m} \frac{\partial^2}{\partial x^2} \right] \Psi$$

b). Schrodinger time independent wave equation

$$\frac{\partial^2 \Psi}{\partial x^2} + \frac{2m}{\hbar^2}(E - V)\Psi = 0$$

13. List the applications of Schrodinger wave equation.

The applications of Schroedinger wave equation

a). particle in one dimensional potential box.

b). describes the wave nature of a particle in mathematical form

14. For a free particle moving within a one dimensional potential box, the ground state energy cannot be zero. Why?

For a free particle moving within a one dimensional potential box, when n = 0, the wave function is zero for all values of x, i.e., it is zero even within the potential box. This would mean that the particle is not present within the box. Therefore the state with n = 0 is not allowed. As energy is proportional to n_2, the ground state energy cannot be zero since n = 0 is not allowed.

Quantum Physics

15. What is wave particle duality?

When the particles are in motion, they exhibit wave nature. This phenomena is known as wave particle duality.

16. What is meant by wave function?

A variable quantity that characterizes matter waves is known as wave function and is denoted by the symbol 'ψ'. It gives the relation between particle and wave nature of the matter.

17. Give the significance of wave function.

a). It gives the relation between the particle and wave.

b). It gives the information about the particle.

c). ψ is a complex quantity and does not have any physical meaning individually.

d). The magnitude of ψ is real and positive.

e). For a given volume dτ, the probability of finding the particle is given by probability (P) = $\iiint |\Psi|^2 dxdydz = 1$

18. What are Eigen values and Eigen function?

Energy of a particle moving in one dimensional potential box of width 'l' is given by

$$E_n = \frac{n^2 h^2}{8ml^2}$$

Each value of E_n is called an eigen value.

Energy eigen function of a particle in one dimensional potential box is given by

$$\Psi_n = \sqrt{\frac{2}{l}} \sin\left(\frac{n\pi x}{l}\right)$$

19. Define tunneling. List out the major applications of quantum tunneling.

It can be defined as that there is a finite chance for the particle to penetrate the potential barrier.

Applications:

a). It can be used as the switches by varying the height of the barrier.

b). It can be used to study the surface of atoms by producing the images.

c). It acts as the tunnel diode in which electrons tunnel through a potential barrier.

20. **What is the principle of scanning tunneling microscope?**

    The principle of scanning tunneling microscope is tunneling effect. Tunneling can be defined as that there is a finite chance for the particle to penetrate the potential barrier.

21. **List out the applications of scanning tunneling microscope.**
    a). It shows the position of atoms
    b). It is used to manipulate the atoms
    c). It is used to analyze the electronic structures of the active sites in a catalyst surfaces

22. **List out the disadvantages of scanning tunneling microscope.**
    a). It is highly sensitive
    b). The presence of single dust particle leads to damage the needle.
    c). A small vibration smashes the tip and the sample together.

PART-B

1. Write the postulates of Planck's quantum theory of radiation. Using quantum theory, derive an expression for the average energy emitted by a black body and arrive at Planck's radiation law in terms of frequency.
2. (i) Derive the expression for Planck's quantum theory of radiation. (ii) Deduce Wien's displacement law & Rayleigh-Jeans law from Planck's quantum theory of radiation.
3. Explain Compton effect and its physical significance. Derive the relation giving the change of wavelength the energy of recoil electron and recoil angle.
4. i) Explain the physical significance of wave function. (ii) What are matter waves? Write the properties of matter waves.
5. Derive Schrodinger's time dependent and time independent wave equations.
6. Derive Schrodinger's wave equation for a particle in a one dimensional box. Solve it to obtain Eigen function and show that Eigen values are discrete.
7. What is meant tunneling?. Explain the construction and working of Scanning Tunneling Microscope.

UNIT-V
CRYSTAL PHYSICS

5.1 Crystalline materials
The materials in which the atoms arranged orderly with regular pattern are called as crystalline materials.

5.1.1 Single crystalline materials
The materials which consist of only one crystal is called as single crystal.

5.1.2 Polycrystalline materials
The materials which made by a collection of many small crystals is known as polycrystalline materials.

5.2 Non-crystalline materials or amorphous materials
Non-crystalline materials are known as that the materials in which the atoms are not arranged orderly with regular pattern.

Space lattice
It is an array of points in space to represent atoms in a crystal.

5.3 Single crystals
5.3.1 Unit cell
The unit cell is defined as the smallest geometric figure which is repeated in three dimensions to derive the actual crystal structure.

5.3.2 Lattice parameters of the unit cell
- **a). Intercepts:** The length of unit cell along X, Y and Z axes is known as intercepts.
- **b). Interfacial angles:** The angles between three intercepts are called interfacial angles.
- **c). Primitive cell:** It contains only one lattice points per unit cell. Ex: Simple cubic.
- **d). Non-primitive cell:** It contains more than one lattice point per unit cell. Ex: BCC & FCC.

5.3.3 Crystals systems

Sl. No	Crystal system	Axial length	Interfacial angle	Examples
1.	Cubic	$a=b=c$	$\alpha=\beta=\gamma=90°$	NaCl
2.	Tetragonal	$a=b\neq c$	$\alpha=\beta=\gamma=90°$	Indium
3.	Orthorhombic	$a\neq b\neq c$	$\alpha=\beta=\gamma=90°$	Sulphur
4.	Monoclinic	$a\neq b\neq c$	$\alpha=\beta=90°, \gamma\neq90°$	Na2SO4
5.	Triclinic	$a\neq b\neq c$	$\alpha\neq\beta\neq\gamma\neq90°$	CuSO4
6.	Rhombohedral	$a=b=c$	$\alpha=\beta=\gamma\neq90°$	Calcite
7.	Hexagonal	$a=b\neq c$	$\alpha=\beta=90°, \gamma=120°$	Zinc

5.3.4 Bravais lattice

There are only 14 possible ways of arranging lattice points in space. These 14 space lattices are called as Bravais lattice.

Sl. No	Crystal systems	No	Bravais lattices
1.	Cubic	3	a) Simple cubic b) Body Centered Cubic c) Face Centered Cubic
2.	Tetragonal	2	a) Simple tetragonal b) Body Centered Tetragonal
3.	Orthorhombic	4	a) Simple Orthorhombic b) Base Centered Orthorhombic c) Body Centered Orthorhombic d) Face Centered Orthorhombic
4.	Monoclinic	2	a) Simple Monoclinic b) Base Centered Monoclinic
5.	Triclinic	1	Simple Triclinic
6.	Rhombohedral	1	Simple Rhombohedral
7.	Hexagonal	1	Simple Hexagonal

5.3.5 Directions and planes in crystals

Directions in crystals: The procedure of finding the directions inside the crystal is explained below.

a). Consider any lattice point that lies on the line as origin.

b). Write down the position vector of the next nearest point on the line in terms of fundamental translation vector $a, \bar{b}$ and $\bar{c}$ of the unit cell of the crystal.

c). Now the components of position vector r along the three directions of a, b, c are r_1, r_2, r_3 respectively. Then the crystal direction is obtained by $[r_1 r_2 r_3]$.

Planes in crystals: A crystal lattice is considered as a collection of a set of parallel equidistant planes passing through lattice points. These planes are known as lattice planes. It is illustrated in Fig. 5.1.

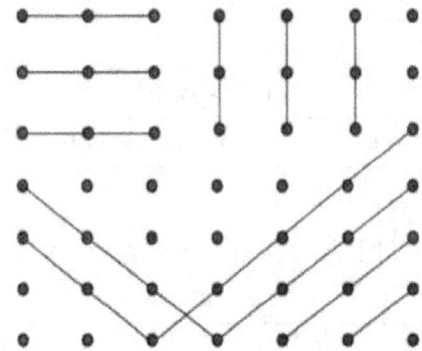

Fig. 5.1 Directions and planes in crystals

5.4 Miller indices

Miller indices can be defined as a set of three numbers to designate a plane in a crystal. This set of three numbers is known as miller indices and represented by (hkl).

Procedure for determining the Miller indices
 a). Find the intercepts made by the plane.
 b). Find the coefficients of the intercepts
 c). Find the reciprocal of the coefficients
 d). Convert these reciprocals into whole numbers
 e). Enclose these whole numbers in a bracket like ( ).

5.5 Interplanar distance or d-spacing

It can be defined as the perpendicular distance between any two successive planes.

Derivation
 a). Consider a plane ABC in cubic crystal with 'a' as length of the cube edge
 b). A normal ON is drawn to the plane ABC from the origin of the cube. Let ON be interplanar spacing (d). i.e. ON = d.
 c). Intercepts are OA, OB and OC.

d). Crystallographic axes are OX, OY and OZ.

e). α′, β′ and γ′ are the angles between crystallographic axes of OX, OY and OZ. They are normal to ON.

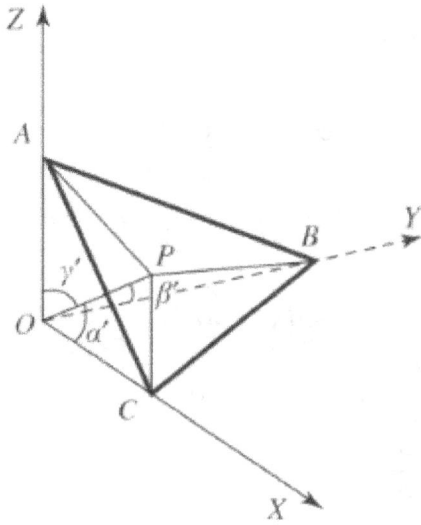

Fig. 5.2 A plane in cubic crystal

We know that intercepts are reciprocal of Miller indices. Therefore,

$$OA : OB : OC = \frac{1}{h} : \frac{1}{k} : \frac{1}{l}$$

Multiplying 'a', we have

$$OA : OB : OC = \frac{a}{h} : \frac{a}{k} : \frac{a}{l}$$

$$\therefore OA = \frac{a}{h}, OB = \frac{a}{k}, OC = \frac{a}{l}$$

In right angle Δ OAN,

$$\cos\alpha' = \frac{ON}{OA} = \frac{d}{a/h} = \frac{hd}{a} \tag{1}$$

In right angle Δ OBN,

$$\cos\beta' = \frac{ON}{OB} = \frac{d}{a/k} = \frac{kd}{a} \tag{2}$$

In right angle Δ OCN,

$$\cos\gamma' = \frac{ON}{OC} = \frac{d}{a/l} = \frac{ld}{a} \qquad (3)$$

From the law of direct cosine,

$$\cos^2\alpha' + \cos^2\beta' + \cos^2\gamma' = 1 \qquad (4)$$

Substituting eqs (1), (2) and (3) in eq (4), we get

$$\left(\frac{d}{a}\right)^2 + \left(\frac{kd}{a}\right)^2 + \left(\frac{ld}{a}\right)^2 = 1$$

$$\frac{d^2}{a^2}(\,^2 + k^2 + l^2) = 1$$

$$\therefore d = \frac{a}{\sqrt{\,^2 + k^2 + l^2}} \qquad (5)$$

Equation (5) represents the d-spacing or interplanar distance or interplanar spacing of an unit cell in cubic.

5.6 Simple cubic structure

The unit cell of simple cubic structure is shown in Fig. 5.3.

a). No atoms per unit cell:

It has 8 corner atoms. Each corner atom is shared by 8 unit cells

$$\therefore \text{No of atoms per unit cell} = \frac{1}{8} \times 8 = 1$$

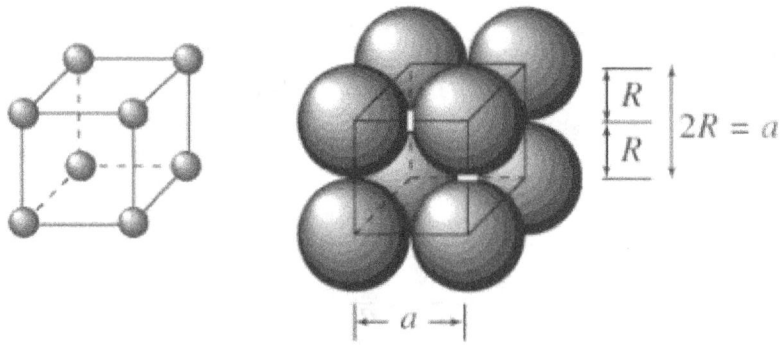

Fig. 5.3 Cubic structure

Crystal Physics

b). Coordination number:

As shown in Fig. 5.4, the reference atom 'x' has 6 nearest neighbouring atoms. So the coordination number is 6.

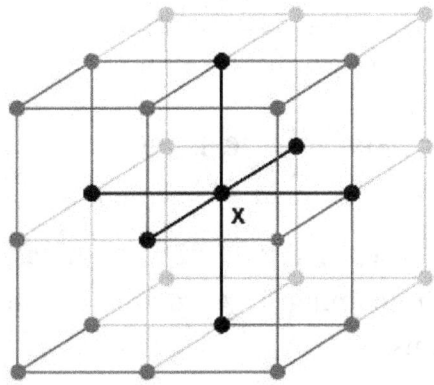

Fig. 5.4

c). Atomic radius:

As shown in Fig. 5.5, corner atom touches each other along the edges. So $a = 2r$.

$$\text{Atomic radius (r)} = \frac{a}{2}$$

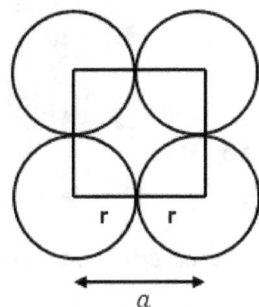

Fig. 5.5

d). Atomic packing factor (APF):

$$APF = \frac{\text{No of atoms per unit cell} \times \text{Volume of one atom}}{\text{Volume of an unit cell}}$$

$$APF = \frac{1 \times \frac{4}{3}\pi r^3}{a^3} = \frac{\frac{4}{3}\pi r^3}{a^3}$$

Crystal Physics

Substituting $r = \dfrac{a}{2}$ in the above equation, we get

$$APF = \dfrac{\dfrac{4}{3}\pi \left(\dfrac{a}{2}\right)^3}{a^3}$$

$$APF = \dfrac{\pi}{6} = 0.52$$

∴ APF% = 52%. It represents that 52% volume of an unit cell is occupied by the atoms and remaining 48% volume of unit cell is vacant or empty or not filled by the atoms.

5.7 Body centered cubic structure (BCC)

An unit cell of BCC has one atom at each corner and one at body center of the cube as shown in the Fig. 5.6.

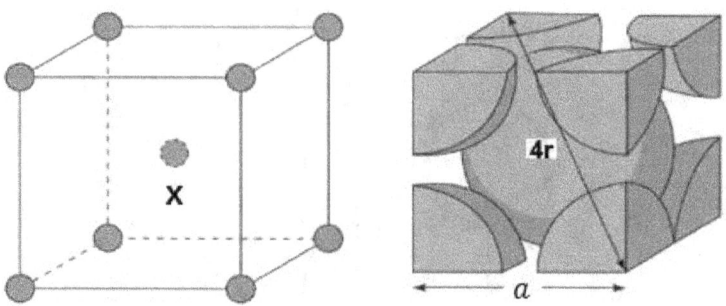

Fig. 5.6 Body centred cubic structure

a). No of atoms per unit cell:

In this unit cell,

    No of corner atoms per unit cell = 1

    No of body centered atoms per unit cell = 1

Total no of atoms per unit cell

= no of corner atoms per unit cell + no of body centered atoms per unit cell

= 1 + 1

= 2.

b). Coordination number:

The reference atom 'x' has 8 nearest neighbouring atoms as shown in Fig. 5.6. So the coordination number is 8

c). Atomic radius

As shown in the Fig. 5.7, consider the atoms A, G and at the centre of the cell 'O'. These atoms lie in a straight line along the body AG of the cube.

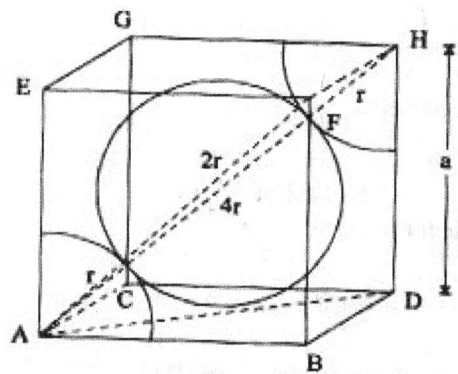

Fig. 5.7

In right angle $\triangle ABC$,

$$AC^2 = AB^2 + BC^2$$

$$AC^2 = a^2 + a^2 = 2a^2$$

In right angle $\triangle ACG$,

$$AG^2 = AC^2 + CG^2$$

$$(r + 2r + r)^2 = 2a^2 + a^2$$

$$16r^2 = 3a^2$$

$$\therefore \text{Atomic radius, } r = \frac{\sqrt{3}a}{4}$$

d) Atomic packing factor (APF):

$$APF = \frac{\text{No of atoms per unit cell} \times \text{Volume of one atom}}{\text{Volume of an unit cell}}$$

$$APF = \frac{2 \times \frac{4}{3}\pi r^3}{a^3}$$

Substituting $r = \frac{\sqrt{3}a}{4}$ in the above equation, we get

$$APF = \frac{\frac{4}{3}\pi \left(\frac{\sqrt{3}a}{4}\right)^3}{a^3}$$

$$APF = \frac{\sqrt{3}\pi}{8} = 0.68$$

∴ APF% = 68%. It represents that 68% volume of an unit cell is occupied by the atoms and remaining 32% volume of unit cell is vacant or empty or not filled by the atoms.

5.8 Face centered cubic structure

An unit cell of FCC has one atom at each corner and one atom at the centre of each face as shown in Fig. 5.8.

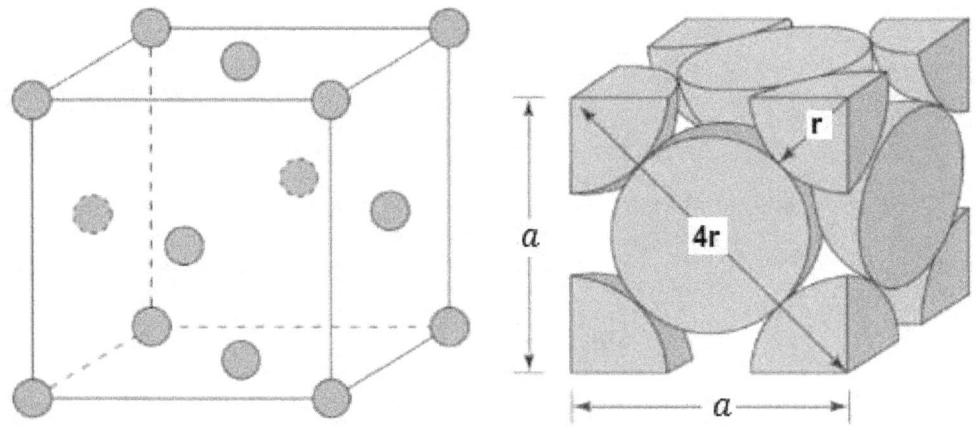

Fig. 5.8 Face centered cubic structure

a) No of atoms per unit cell:

In this unit cell,

    No of corner atoms per unit cell = 1

    No of corner face centered atoms per unit cell = 3

Total no of atoms per unit cell

= no of corner atoms per unit cell + no of face centered atoms per unit cell

= 1+3

Total no of corner atoms = 4.

b) Coordination number:

The reference atom 'x' has 12 nearest neighbouring atoms as shown in Fig. 5.9. So the coordination number is 12.

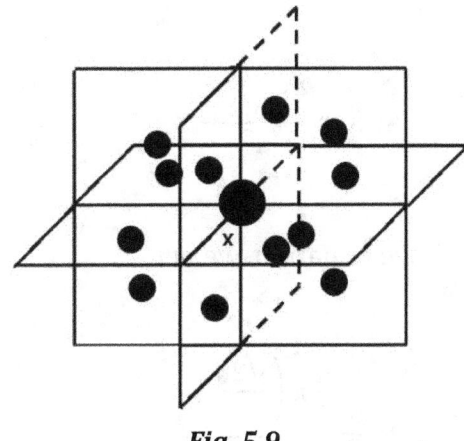

Fig. 5.9

c) Atomic radius:

As shown in Fig. 5.10, consider the atoms at A, C and at the centre of face 'O'. These atoms lie in a straight line along the body AC of the cube.

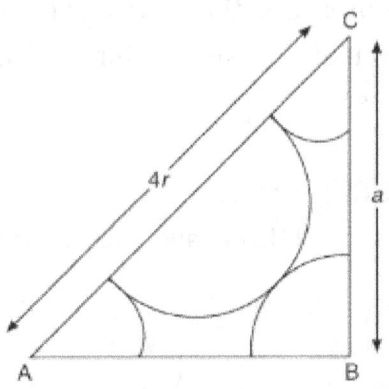

Fig. 5.10

In right angle $\triangle ABC$,

$$AC^2 = AB^2 + BC^2$$

$$(r + 2r + r)^2 = a^2 + a^2$$

$$16r^2 = 2a^2$$

$$r = \frac{\sqrt{2}a}{4}$$

d) Atomic packing factor (APF):

$$APF = \frac{\text{No of atoms per unit cell} \times \text{Volume of one atom}}{\text{Volume of an unit cell}}$$

$$APF = \frac{4 \times \frac{4}{3}\pi r^3}{a^3}$$

Substituting $r = \frac{\sqrt{2}a}{4}$ in the above equation, we get

$$APF = \frac{\frac{4}{3}\pi\left(\frac{\sqrt{2}a}{4}\right)^3}{a^3}$$

$$APF = \frac{\sqrt{2}\pi}{6} = 0.74$$

∴ APF% = 74%. It represents that 74% volume of an unit cell is occupied by the atoms and remaining 26% volume of unit cell is vacant or empty or not filled by the atoms.

Hexagonal closely packed (HCP) structure

An unit cell of HCP consists of three layers as shown in Fig. 5.11.

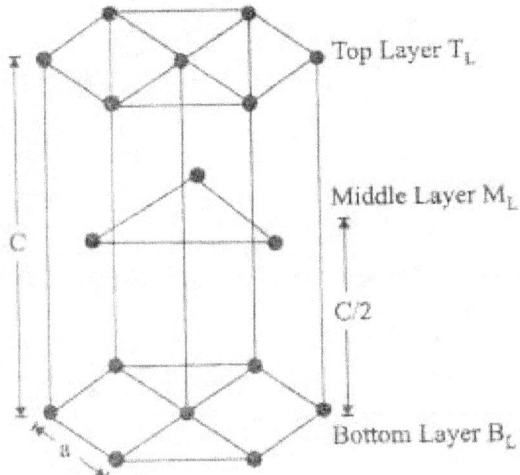

Fig. 5.11 HCP structure

Crystal Physics

(i). Bottom layer: The central atom has 6 nearest neighboring atoms

(ii). Middle layer: It has three atoms

(iii). Top layer: The central atom has 6 nearest neighboring atoms

a). No of atoms per unit cell:

In this unit cell,

No of corner atoms per unit cell = 2

No of base central atoms per unit cell = 1

No of middle layer atoms per unit cell = 3

Total no of atoms per unit cell

= (No of corner atoms per unit cell+ No of base central atoms per unit cell+ No of middle layer atoms per unit cell)

= 2+1+3

Total no of atoms per unit cell = 6 atoms.

b). Coordination number:

As shown in the Fig. 5.12, the reference atom 'x' has 6 neighboring atoms in its own plane. It has 3 atoms in its above layer and 3 more atoms below the layer. So coordination number = 6+3+3 =12.

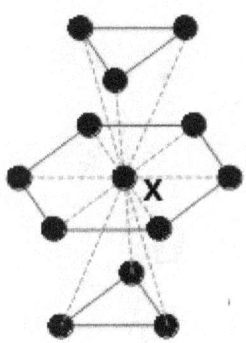

Fig. 5.12

c). Atomic radius:

As shown in Fig. 5.13, each corner atom touches each other. 'a' be the length of the side and 'r' be the radius of the atoms. So $a = 2r$.

$$\therefore \text{Atomic radius (r)} = \frac{a}{2}$$

Crystal Physics

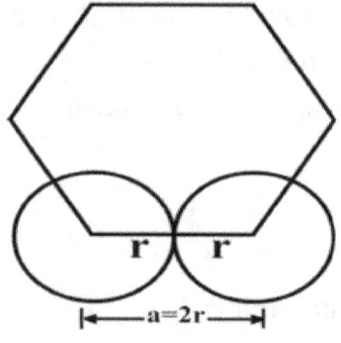

Fig. 5.13

Calculation of c/a ratio:

Consider a triangle ABO in the bottom layer of HCP as shown in Fig. 5.14. Let R be the mid-point of AB and join OR.

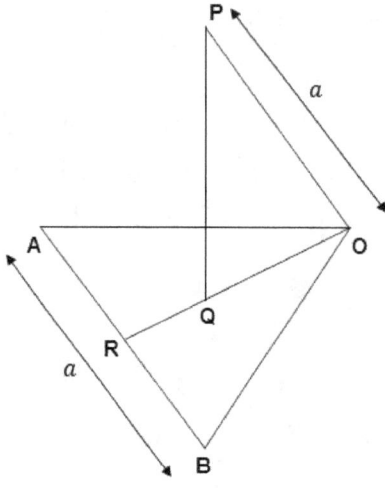

Fig. 5.14

In $\triangle$ AOR,

$$\cos 30° = \frac{OR}{OA}$$

$$OR = OA\cos 30° = \frac{a\sqrt{3}}{2} \tag{1}$$

If Q is the orthocenter of $\triangle OAB$, then

$$OQ = \frac{2}{3} OR \tag{2}$$

substituting eq(1) in eq(2), $OQ = \dfrac{2}{3}\dfrac{a\sqrt{3}}{2} = \dfrac{a}{\sqrt{3}}$ (3)

In $\triangle OPQ$,
$$OP^2 = OQ^2 + PQ^2 \qquad (4)$$

Substituting $OP = a$, $OQ = \dfrac{a}{\sqrt{3}}$ and $PQ = \dfrac{c}{2}$ in eq (4), we get

$$a^2 = \left(\dfrac{a}{\sqrt{3}}\right)^2 + \left(\dfrac{c}{2}\right)^2$$

$$a^2 = \dfrac{a^2}{3} + \dfrac{c^2}{4}$$

$$a^2 - \dfrac{a^2}{3} = \dfrac{c^2}{4}$$

$$\dfrac{2a^2}{3} = \dfrac{c^2}{4}$$

$$\dfrac{c^2}{a^2} = \dfrac{8}{3}$$

$$\dfrac{c}{a} = \sqrt{\dfrac{8}{3}} = 1.633$$

d) Atomic packing factor (APF):

$$APF = \dfrac{\text{No of atoms per unit cell} \times \text{Volume of one atom}}{\text{Volume of an unit cell}}$$

$$APF = \dfrac{6 \times \dfrac{4}{3}\pi r^3}{6 \times \text{base area} \times \text{height}}$$

$$= \dfrac{6 \times \dfrac{4}{3}\pi r^3}{6 \times \dfrac{1}{2} \times OA \times OR \times c}$$

$$APF = \frac{6 \times \frac{4}{3}\pi r^3}{6 \times \frac{1}{2} \times a \times \frac{a\sqrt{3}}{2} \times c} \quad (5)$$

Substituting, $r = \frac{a}{2}$ in eq (5), we get

$$APF = \frac{6 \times \frac{4}{3}\pi \left(\frac{a}{2}\right)^3}{6 \times \frac{1}{2} \times a \times \frac{a\sqrt{3}}{2} \times c}$$

$$= \frac{\frac{4}{3}\pi \frac{a^3}{8}}{\frac{a^2\sqrt{3}c}{4}} = \frac{\frac{\pi a^3}{6}}{\frac{a^2\sqrt{3}c}{4}}$$

$$APF = \frac{\pi a^3}{\frac{3\sqrt{3}a^2 c}{2}} = \frac{\pi}{3\sqrt{2}}$$

$$APF = 0.74$$

∴ APF% = 74%. It represents that 74% volume of an unit cell is occupied by the atoms and remaining 26% volume of unit cell is vacant or empty or not filled by the atoms.

5.10 Diamond structure

Diamond is termed due to the combination of two interpenetrating FCC sublattices, having the origin (0,0,0) and $\left(\frac{1}{4}, \frac{1}{4}, \frac{1}{4}\right)$ along the body diagonal. It has three types of atoms namely, corner atom, face centered atoms and four atoms presents inside the unit cell represented as 1, 2, 3, 4 in Fig. 5.15.

a) No of atoms per unit cell:
In this unit cell,

No of corner atoms per unit cell = 1 atom

No of face centered atoms per unit cell = 3 atoms

No of atoms inside the unit cell = 4 atoms

Total no of atoms per unit cell = 1+3+4 = 8 atoms

Crystal Physics

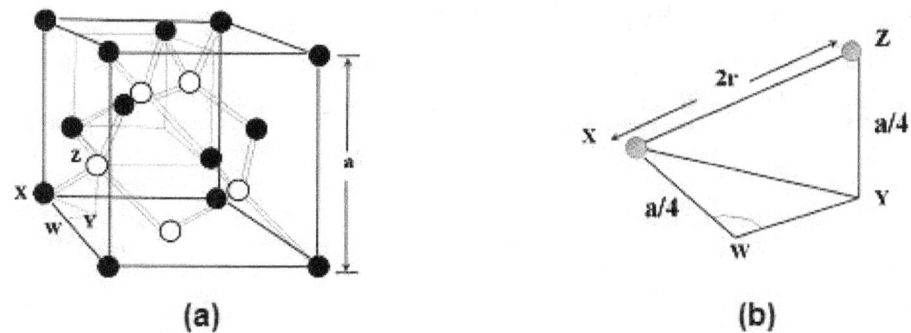

Fig. 5.15 (a) Diamond structure and (b) calculation of atomic radius

b) Coordination number:

As shown in the Fig. 5.15(a), the reference atom 'x' has 4 nearest neighboring atoms. So coordination number = 4.

c) Atomic radius:

In unit cell of diamond structure, corner atoms and face centered atoms do not have contact with each other. Both the face centered atom and corner atom have contact with 4 atoms (1, 2, 3, 4) situated inside the unit cell.

From the Fig. 5.15(b),

$$\text{In } \Delta XYZ, \quad XY^2 = XZ^2 + ZY^2$$

$$XY^2 = [(XT)^2 + (TZ)^2 + (ZY)^2]$$

$$(2r)^2 = \left(\frac{a}{4}\right)^2 + \left(\frac{a}{4}\right)^2 + \left(\frac{a}{4}\right)^2$$

$$4r^2 = \frac{a^2}{16} + \frac{a^2}{16} + \frac{a^2}{16}$$

$$4r^2 = \frac{a^2 + a^2 + a^2}{16}$$

$$4r^2 = \frac{3a^2}{16}$$

$$r^2 = \frac{3a^2}{64}$$

Atomic radius, $\quad r = \dfrac{a\sqrt{3}}{8}$

Crystal Physics

d) Atomic packing factor (APF):

$$APF = \frac{\text{No of atoms per unit cell} \times \text{Volume of one atom}}{\text{Volume of an unit cell}}$$

$$APF = \frac{8 \times \frac{4}{3}\pi r^3}{a^3}$$

Substituting, $r = \frac{a\sqrt{3}}{8}$ in the above equation, we get

$$APF = \frac{8 \times \frac{4}{3}\pi \left(\frac{a\sqrt{3}}{8}\right)^3}{a^3}$$

$$APF = \frac{\sqrt{3}\pi}{16} = 0.34$$

∴ APF% = 34%. It represents that 34% volume of an unit cell is occupied by the atoms and remaining 66% volume of unit cell is vacant or empty or not filled by the atoms.

5.11 Crystal imperfections

The disturbance occurred in the regular orientation of atoms is called defect or imperfections.

Classification of crystal imperfections:
    a). Point defects
    b). Line defects
    c). Surface defects
    d). Volume defects

a). Point defects
1. They are crystalline irregularities.
2. They take place due to imperfect packing of atoms
3. They produce distortion inside the crystal
4. They also produce strain.

Types of point defects
  i). Vacancies
  ii). Interstitial
  iii). Impurities

i) Vacancies:

Whenever one or more atoms are missing from a normally occupied position, the defect caused is known as vacancy.

There are different kinds of vacancies like Frenkel defect, Schottky defect, Colour defect, etc.

Frenkel defects: A vacancy associated with interstitial impurity is called Frenkel defect. It is shown in Fig. 5.16.

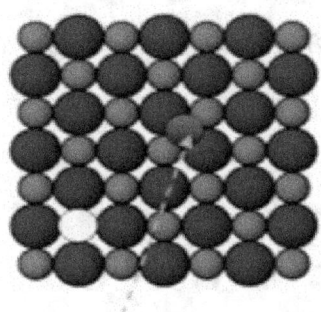

Fig. 5.16 Frenkel defect

Schottky defect: It refers to the missing of a pair of positive and negative ions in an ionic crystal. It is shown in Fig. 5.17

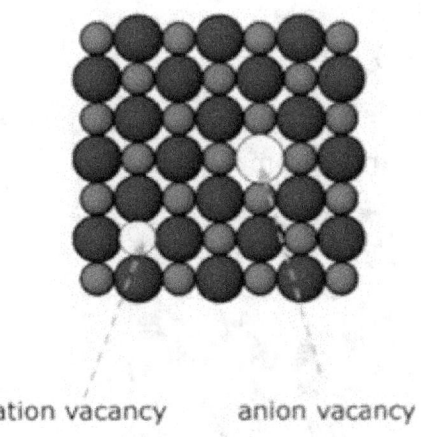

Fig. 5.17 Schottky defect

ii) Interstitial defects:

When an extra atom occupies interstitial space within the crystal structure without removing parent atom, the defect is called interstitial defect. There are two types of interstitial defects:

Self interstitial defect: It is shown in Fig. 5.18. An atom from same crystal occupied in interstitial site is called self interstitial site.

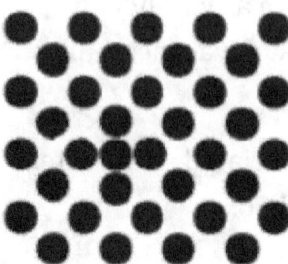

Fig. 5.18 Self interstitial defect

Foreign interstitial defect: It is shown in Fig. 5.19. An impurity atom occupied in interstitial site is called foreign interstitial defect.

Fig. 5.19 Foreign interstitial effect

iii) Impurities:

When the foreign atoms (impurities) are added to crystal lattices, the defect is called impurity defects.

There are two types of impurity defects

1. Substitutional impurity defect: It is shown in Fig. 5.20 and refers to a foreign atom that replaces a parent atom in the lattice.

Fig. 5.20 Substitutional impurity defect

2. Interstitial impurity defect: It refers that the small size atom occupying the empty space (interstitial) in the parent crystal. It is shown in Fig. 5.21.

Fig. 5.21 Interstitial impurity defect

b). Line defects

The defect due to dislocation or distortion of atoms along a line is known as line defect.

There are two types of line defects

(i) Edge dislocation: It arises when one of the atomic planes forms only partially and does not extend through the entire crystal. Edge dislocation is further divided into two types.

Positive edge dislocation: If the extra plane of atoms is above the slip plane of the crystal, then the edge dislocation is called positive edge dislocation. It is shown in Fig. 5.22.

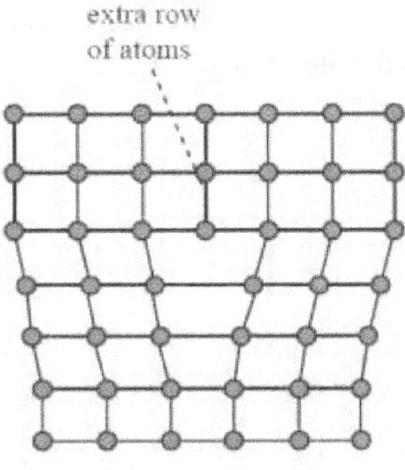

Fig. 5.22 Positive edge dislocation

Crystal Physics

Negative edge dislocation: If the extra plane of atoms is just below the slip plane of the crystal, then the edge dislocation is called positive edge dislocation. It is shown in Fig. 5.23.

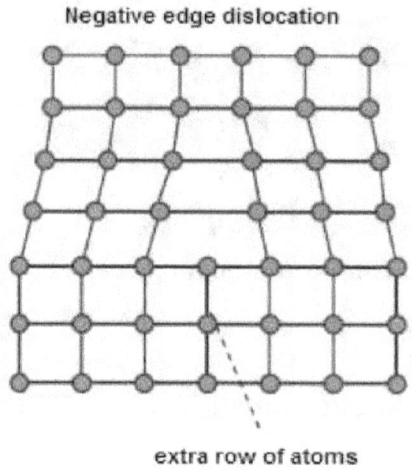

Fig. 5.23 Negative edge dislocation

(ii) **Screw dislocation:** Screw dislocation can be formed in a crystal structure by applying upward and downward shear stress to regions of a perfect crystal which have been separated by a cutting plane. It is shown in Fig. 5.24. In screw dislocation plane of the crystal lattice trace a helical path around the dislocation line.

The burgers vector of screw dislocation is parallel to the dislocation line and when stress is applied on crystal having this defect then the dislocation area moves perpendicular to the direction of stress.

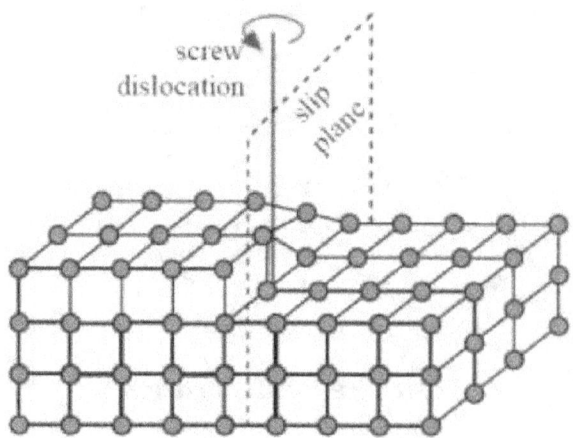

Fig. 5.24 Screw dislocation

Crystal Physics

5.12 Burger's vectors

The presence of dislocation results in lattice distortion. The magnitude and the direction of such distortion is expressed in terms of a vector called Burger's vector. It characterizes a dislocation line and represented as 'b'. For an edge dislocation, b is perpendicular to the dislocation line, whereas in the cases of the screw dislocation it is parallel.

Procedure for finding the Burger's vector of a dislocation:

The Burgers vector associated with a dislocation is a measure of the lattice distortion caused by the presence of the line defect. Fig. 5.25 shows the convention for measuring the Burgers vector. A circuit is made around a dislocation line in a clockwise direction (top picture) with each step of the circuit connecting lattice sites that are fully coordinated. This circuit is then transferred to a perfect lattice of the same type. Because of the absence of a dislocation within this circuit, it fails to close on itself, and the vector linking the end of the circuit to the starting point is the Burgers vector, **b** = QM.

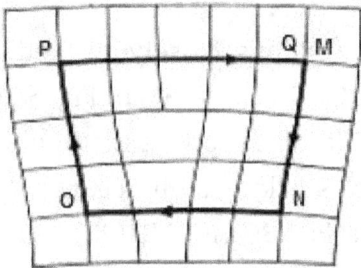

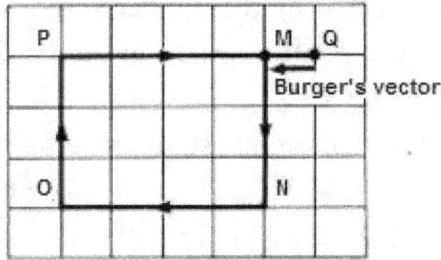

Fig. 5.25 Burger vectors of a dislocation

The Burgers vector defined in this way is a unit vector of the lattice if the dislocation is a unit dislocation, and a shorter stable translation vector of the lattice if the dislocation is a partial dislocation.

5.13 Stacking faults

Whenever the stacking of atoms is not in proper sequence throughout the crystal, the defect caused is called stacking fault. It is shown in Fig. 5.26.

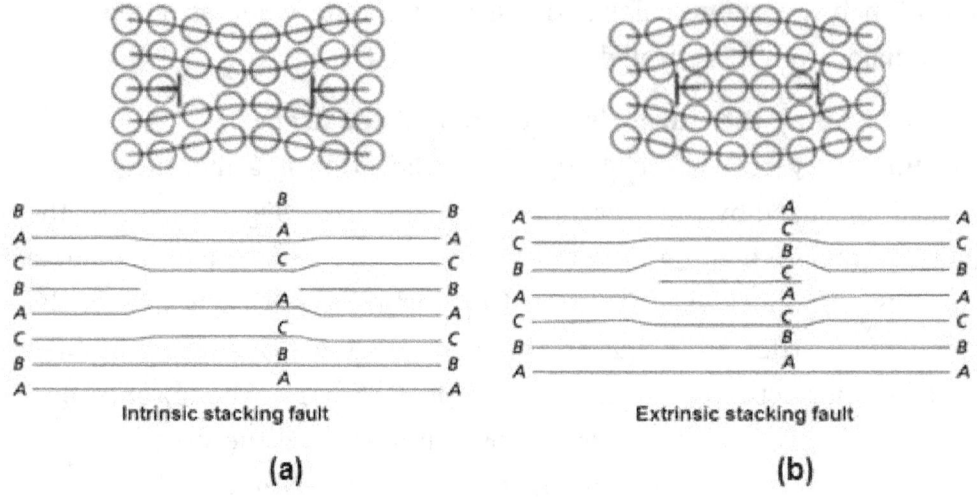

Fig. 5.26 Stacking faults

For instance, in FCC lattice, alloys closely packed atomic layers are normally in the alteration ABC, ABC, ABC as shown in Fig. 5.26(a).

In the faulty stacking(Fig. 5.26(b)), suppose the two C layers are missing in the sequence, the region in which stacking fault occurs as AB, AB.

Now the stacking sequence is ABC AB AB ABC which belongs to HCP structure. Thus the region AB AB is called stacking fault.

5.14 Role of imperfection in plastic deformations

Suppose a crystal is deformed by the application of stresses, it returns to its original state upon removal of stresses. Then the deformation is said to be elastic.

When it does not return to its original state, then the deformation is said to be plastic.

In most of the crystals, the plastic deformations result from the slip of one part of the crystal relative to another. The most important feature of plastic deformation is thus its non-reversibility. From an atomic point of crystal view, plastic deformation belongs to the breaking of bonds with original atom neighbours and then reforming bonds with new neighbours as large numbers of atoms or molecules move relative to one another.

5.15 Growth of single crystals from solution

It is a simplest method to grow crystals which are highly soluble. There are sub divided into following methods.

5.15.1 Solution growth
1. Slow cooling method:

In this method, a saturated solution is taken in a bath. A seed crystal is suspended in the solution and the temperature is reduced at slow rate. The crystallization begins in the temperature range of 45°C - 75°C and the lower limit of cooling is the room temperature. The seed produces a large size of single crystal.

2. Slow evaporation method:

In this method, the super saturation of a solution is achieved by evaporating the solvent at a fixed temperature. The solution concentration increases and grows into crystal on a seed crystal.

3. Temperature gradient method:

The setup of temperature gradient method is shown in Fig. 5.27. It consists of a large tank and a constant temperature bath. The solution is prepared by dissolving the substance in a solvent. The seed crystal is introduced inside the solution and gently rotated by an electric rotator.

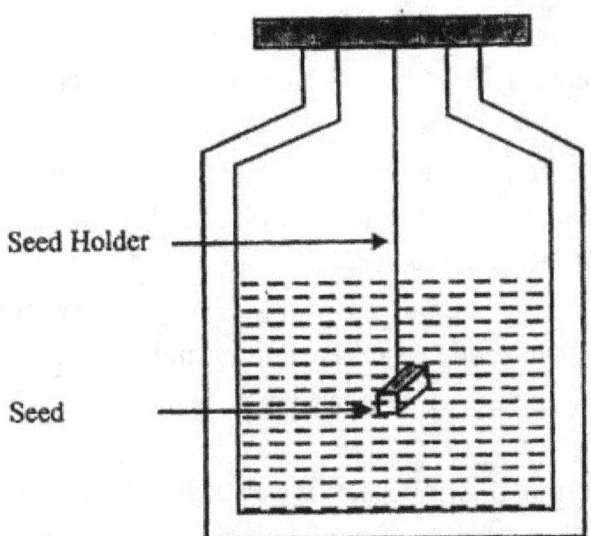

Fig. 5.27 Temperature gradient method

   a). Low temperature (35°C-100°C) solution growth.

   b). High temperature (about 120°C) solution growth.

Advantages
> a). This is a simple and convenient method of growing single crystals
> b). Growth of defect free crystals is possible

Disadvantages
> a). It is applicable only for substances fairly soluble in a solvent
> b). Rate of evaporation may affect the quality of the crystal

5.15.2 Melt growth

It is the process of crystallization by fusion and re-solidification of the starting materials from the melt. There are various techniques available. The main techniques are
> a). Czochralski method
> b). Bridgmann method
> c). Verneuil method
> d). Zone melting method

a) Czochralski method
Principle:

It is the process of crystallization by fusion and re-solidification of the starting materials from the melt.

Construction and working:

The schematic diagram is shown in Fig. 5.28. An apparatus consists of the following components: i) Crucible, ii) Heater, iii) Seed crystal, iv) Crystal holder.

The material is taken in the crucible. The material is heated above the melting point using heater. Thus, the melt is obtained in the crucible. A seed crystal is introduced into the melt using crystal holder.

A small portion of seed crystal is initially melted. The temperature is then suitably adjusted. The seed crystal is rotated and gradually pulled out of the melt by maintaining the grown crystal. Thus, a single crystal grows on the seed crystal.

The diameter of the grown crystal is controlled by the temperature of the melt and rate of pulling. The seed crystal act as a nucleation centre i.e., the solidified material at the surface of the seed will reproduce single-crystal structure of seed.

Crystal Physics

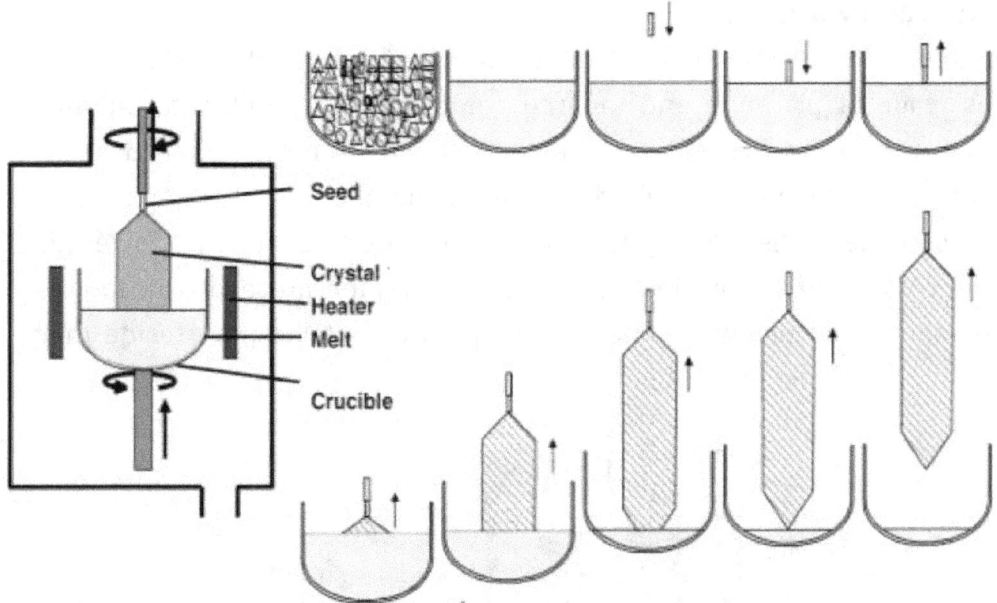

Fig. 5.28 Crystal growth from Czochralski method

Advantages:
   a). It can produce large single crystal
   b). Defect free crystal is possible
   c). It allows convenient chemical composition of crystal.

Disadvantages:
   a). High vapour pressure of the materials can be produced.
   b). It may produce contamination of melt by the crucible.

b) Bridgmann method
Principle

It is based on directional solidifications by translating a molten material from the hot to the cold zone of the furnace.

There are two types of Bridgman methods.
   a). Vertical Bridgman technique
   b). Horizontal Bridgman technique

In both techniques, the melt in a sealed crucible is frozen from one end by one of the following methods.
   a). Moving the crucible from high temperature region to low temperature region of the furnace.
   b). Moving the furnace over crucible
   c). Both furnace and crucible are stationary.

Construction and working:

In this method, the material is taken in a cylindrical crucible. The crucible is made of platinum and tapered conically with pointed tip at the bottom as shown in Fig. 5.29. The crucible is suspended in the upper furnace until the material is completely melted into molten liquid.

The crucible is then slowly lowered from upper furnace into lowered furnace with the help of an electric motor. The temperature of the lower furnace is maintained below the melting temperature of the melt inside the crucible.

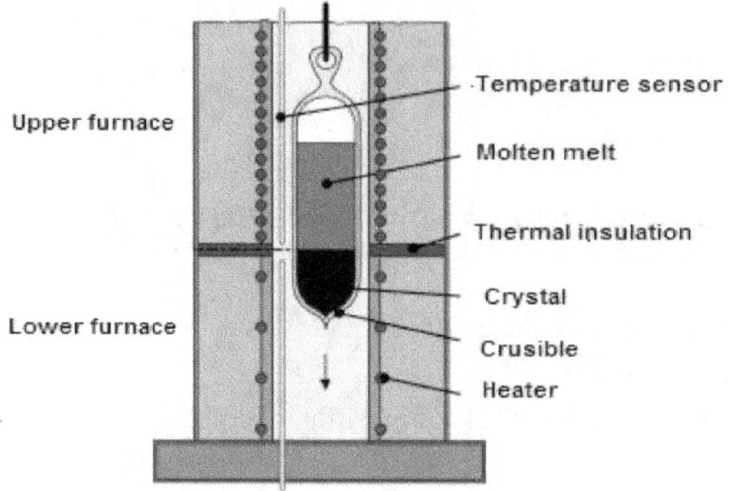

Fig. 5.29 Crystal growth from Bridgmann method

When the pointed tip entered into the lower furnace, the melt starts to solidify to form crystal. As the crucible is continuously lowered, the solidification of melt continuous to form crystal until all the melt becomes solid crystal.

Advantages:
a). Control over vapor pressure is possible during growth
b). Oxidation of melt is prevented
c). It enables the easy stabilization of temperature gradients.

Disadvantages:
a). Crystal perfection is not better than that of the seed
b). No visibility of material during growth

Crystal Physics

PART-A TWO MARKS

1. What are single crystalline materials and polycrystalline materials?

A material which contains only one crystal is known as single crystal. A material which made by a collection of many small crystals is known as polycrystalline materials

2. Distinguish between crystalline and amorphous materials.

Sl. No	Crystalline materials	Amorphous Materials
1.	Atoms or molecules are arranged in an orderly fashion	Atoms or molecules are not arranged in orderly fashion.
2.	They are anisotropic	They are isotrophic
3.	They have sharp melting point	They do not have sharp melting point
4.	Examples: Diamond, Nacl	Examples: Glass, plastic

3. What is meant by primitive and non-primitive cell? Give an example.

Primitive cell which contains only one lattice point per unit cell. Example : Simple Cubic (SC). Non-primitive cell which contains more than one lattice point per unit cell. Example: BCC & FCC

4. Define unit cell.

An unit cell can be defined as the smallest geometric figure which is repeated in three dimensions to derive the actual crystal structure.

5. What are Bravais lattice?

There are only 14 ways of arranging points in space such that the environment looks same from each point. i.e., there are 14 possible types of space lattices out of the seven crystal systems. These 14 space lattices are called as Bravais lattices.

6. Define lattice plane.

A crystal lattice is considered as a collection of a set of parallel equidistant planes passing through lattice points. These planes are known as lattice planes.

7. What are lattice parameters for a unit cell?

The intercepts on the axes a, b and c and interfacial angles α, β and γ are called lattice parameters of a unit cell.

Crystal Physics

8. **Define space lattice. How it is useful to describe a crystal structure?**

   It is an array of points in space to represent atoms in a crystal.

9. **What are Miller indices? Explain the procedure for determining the miller indices.**

   Miller indices are defined as the reciprocal of the integers made by the plane on the crystallographic axes which are reduced to smallest numbers.

   Procedure for determining the miller indices
   a). Find the intercepts made by the plane
   b). Find the coefficients of the intercepts
   c). Find the reciprocal of the coefficients
   d). Convert these reciprocals into whole numbers
   e). Enclose these whole numbers in a bracket like (hkl)

10. **Define interplanar distance or d-spacing.**

    It can be defined as the perpendicular distance between any two successive planes.

11. **For a cubic system, sketch the planes with Miller Indices (101), (110) and (011).**

    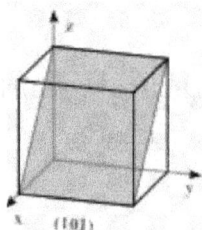
    (101)
    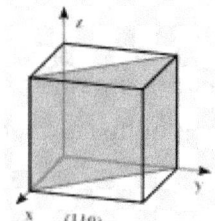
    (110)
    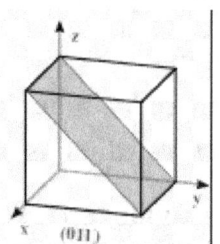
    (011)

12. **A crystal plane cuts at 3a, 4b and 2c distances along the crystallographic axes. Find the Miller Indices of the plane.**

    Intercepts: 3a, 4b, 2c

    Coefficients of intercepts: 3, 4, 2

    Reciprocal of coefficients $\frac{1}{3}, \frac{1}{4}, \frac{1}{2}$

    LCM of denominators 3, 4 and 2 is 12.

    Hence, $\frac{1}{3} \times 12 = 4$, $\frac{1}{4} \times 12 = 3$, $\frac{1}{2} \times 12 = 6$

    Miller induces of the plane is **(436)**.

Crystal Physics

13. Determine the lattice constant for FCC Lead crystal of radius 1.746 Å.

Given: Atomic radius: 1.746 Å = 1.746×10⁻¹⁰ m.

$$r = \frac{\sqrt{2}}{4}a$$

$$\text{Lattice constant, } a = \frac{4r}{\sqrt{2}} = \frac{4 \times 1.746 \times 10^{-10}}{\sqrt{2}}$$

$$= \frac{4 \times 1.746 \times 10^{-10}}{1.414}$$

$a = 4.94 \times 10^{-10}$ m.

14. The lattice constant for an unit cell of aluminium is 4.049 Å. Calculate the spacing of (220) plane.

Given: a = 4.049 Å; h = 2, k = 2, l = 0.

$$\text{d spacing, } d = \frac{a}{\sqrt{h^2 + k^2 + l^2}} = \frac{4.049 \times 10^{-10}}{\sqrt{2^2 + 2^2 + 0^2}} = \frac{4.049 \times 10^{-10}}{\sqrt{8}}$$

$$d = 1.432 \times 10^{-10} \text{ m}$$

15. Define atomic packing fraction.

The ratio between the volume occupied by the total number of atoms per unit cell (v) to the total volume of the unit cell (V) is called atomic packing facto (APF).

16. How carbon atoms are arranged in diamond structure?

Diamond is formed due to the combination of two interpenetrating FCC sublattices, having the origin (0, 0, 0) and (¼, ¼, ¼) along the body diagonal. It has three types of atoms namely corner atoms, face centered atoms and four atoms presents inside the unit cell represented as 1, 2, 3, 4.

17. What is meant by crystal imperfection.

The disturbance occurred in the regular orientation of atoms is called crystal defect or imperfection.

18. List out the types of crystal imperfections
   a). Point defects
   b). Line defects

c). Surface defects

d). Volume defects

19. What are Schottky defect and interstitial defect?

Schottky defects refer to the missing of a pair of positive and negative ions in an ionic crystal. Interstitial defects refer to a vacancy associated with interstitial impurity.

20. Define line defects.

The defects due to the dislocation or distortion of atoms along a line are called as line defects.

21. Define (i) edge dislocation and (ii) screw dislocation

(i) Edge dislocation: It arises when one of the atomic planes forms only partially and does not extend through the entire crystal.

(ii) Screw dislocation: It is due to a displacement of atoms in one part of a crystal relative to rest of the crystal.

22. Define stacking faults.

Whenever the stacking of atoms is not in proper sequence throughout the crystal, the defect caused is called stacking fault.

23. Define Burger vector.

The magnitude and the direction of the displacement of lines are defined by a vector called Burgers vector which characterizes a dislocation line.

24. What is the cause of plastic deformation?

When an applied stress is removed, a crystal does not regain its original shape and size. The deformation is said to be plastic deformation.

25. Name few techniques of crystal growth from melt.

The two important melt growth techniques are (i) Czochralski technique and (ii) Bridgman technique

PART-B

1. Derive the expression for the inter-planar spacing or d-spacing for (hkl) planes of a cubic structure.
2. Explain the no. of atoms, atomic radius, co-ordination number and packing factor for SC, BCC and FCC structures.

Crystal Physics

3. Explain the no. of atoms, atomic radius, co-ordination number and packing factor for HCP structure.
4. Show that the packing factor of FCC and HCP are equal.
5. Explain diamond cubic structure and obtain its no.of atoms per unit cell, atomic radius, co-ordination number and atomic packing factor.
6. What is meant by crystal defects? Explain the various types of crystal defects with neat diagram.
7. Explain the role of imperfections in plastic deformation.
8. Explain the various solution growth techniques along with its merits and demerits.
9. Explain principle, construction and working of Czochralski's method.
10. Explain principle, construction and working of Bridgmann technique

www.ingramcontent.com/pod-product-compliance
Lightning Source LLC
Chambersburg PA
CBHW080545220526
45466CB00010B/3035